MEDICAL
INTELLIGENCE
UNIT

Off Pump Coronary Artery Bypass Surgery

Raymond Cartier, M.D.

Professor of Surgery
University of Montreal
Department of Cardiac Surgery
Montreal Heart Institute
Montreal, Québec, Canada

CRC Press
Taylor & Francis Group
Boca Raton London New York

CRC Press is an imprint of the
Taylor & Francis Group, an **informa** business

OFF PUMP CORONARY ARTERY BYPASS SURGERY

Medical Intelligence Unit

First published 2005 by Landes Bioscience

Published 2018 by CRC Press
Taylor & Francis Group
6000 Broken Sound Parkway NW, Suite 300
Boca Raton, FL 33487-2742

© 2005 by Taylor & Francis Group, LLC
CRC Press is an imprint of Taylor & franc is Group, an Informa business

First issued in paperback 2019

No claim to original U.S. Government works

ISBN 13: 978-0-367-44651-2 (pbk)
ISBN 13: 978-1-58706-075-5 (hbk)

This book contains information obtained from authentic and highly regarded sources. While all reasonable efforts have been made to publish reliable data and information, neither the author[s] nor the publisher can accept any legal responsibility or liability for any errors or omissions that may be made. The publishers wish to make clear that any views or opinions expressed in this book by individual editors, authors or contributors are personal to them and do not necessarily reflect the views/opinions of the publishers. The information or guidance contained in this book is intended for use by medical, scientific or health-care professionals and is provided strictly as a supplement to the medical or other professional's own judgement, their knowledge of the patient's medical history, relevant manufacturer's instructions and the appropriate best practice guidelines. Because of the rapid advances in medical science, any information or advice on dosages, procedures or diagnoses should be independently verified. The reader is strongly urged to consult the relevant national drug formulary and the drug companies' and device or material manufacturers' printed instructions, and their websites, before administering or utilizing any of the drugs, devices or materials mentioned in this book. This book does not indicate whether a particular treatment is appropriate or suitable for a particular individual. Ultimately it is the sole responsibility of the medical professional to make his or her own professional judgements, so as to advise and treat patients appropriately. The authors and publishers have also attempted to trace the copyright holders of all material reproduced in this publication and apologize to copyright holders if permission to publish in this form has not been obtained. If any copyright material has not been acknowledged please write and let us know so we may rectify in any future reprint.

Visit the Taylor & Francis Web site at
http://www.taylorandfrancis.com

and the CRC Press Web site at
http://www.crcpress.com

While the authors, editors and publisher believe that drug selection and dosage and the specifications and usage of equipment and devices, as set forth in this book, are in accord with current recommendations and practice at the time of publication, they make no warranty, expressed or implied, with respect to material described in this book. In view of the ongoing research, equipment development, changes in governmental regulations and the rapid accumulation of information relating to the biomedical sciences, the reader is urged to carefully review and evaluate the information provided herein.

Library of Congress Cataloging-in-Publication Data

Off pump coronary artery bypass surgery / [edited by] Raymond Cartier.
　　　p. ; cm. -- (Medical intelligence unit)
　Includes bibliographical references and index.
　ISBN 1-58706-075-2
1. Coronary artery bypass. I. Cartier, Raymond, 1955- II. Series: Medical intelligence unit (Unnum-bered : 2003)
　　[DNLM: 1. Coronary Artery Bypass. WG 169 O317 2004]
　　RD598.35.C67O37 2004
　　617.4'12--dc22

2004026259

"A mon frère Paul, pour ce qu'il m'a appris et permis d'apprendre..."

"To my brother Paul, for what he taught me and allowed me to learn..."

CONTENTS

EDITOR

Raymond Cartier, M.D.
Professor of Surgery
University of Montreal
Department of Cardiac Surgery
Montreal Heart Institute
Montreal, Québec, Canada
Chapters 1, 3, 4, 6, 8, 10-12

CONTRIBUTORS

Jehangir J. Appoo, M.D.
University of Alberta
Division of Cardiac Surgery
University of Alberta Hospital
Edmonton, Alberta, Canada
Chapter 11

Denis Babin, M.S.C., R.T.
University of Montreal
Department of Anaesthesiology
Montreal Heart Institute
Montreal, Québec, Canada
Chapter 9

Robert Blain, M.D.
Assistant Professor of Anaesthesiology
University of Montreal
Department of Anaesthesiology
Montreal Heart Institute
Montreal, Québec, Canada
Chapter 5

Ray C.-J. Chiu, M.D.
Professor of Surgery
McGill University
Division of Cardiovascular Surgery
Montreal General Hospital
Montreal, Québec, Canada
Chapter 14

Pierre Couture, M.D.
Assistant Professor of Anaesthesiology
University of Montreal
Department of Anaesthesiology
Montreal Heart Institute
Montreal, Québec, Canada
Chapter 9

François Dagenais, M.D.
Assistant Professor of Surgery
Laval University
Division of Cardiac Surgery
Laval Hospital
Québec City, Québec, Canada
Chapter 2

Roland G. Demaria, M.D.
Assistant Professor of Surgery
University of Montpellier
Department of Cardiovascular Surgery
Arnaud de Villeneuve Hospital
Montpellier, France
Chapter 7

André Denault, M.D.
Associate Professor of Anaesthesiology
University of Montreal
Department of Anaesthesiology
Montreal Heart Institute
Montreal, Québec, Canada
Chapter 9

Nicolas Dürrleman, M.D.
Assistant Professor of Surgery
University of Montpellier
Department of Cardiovascular Surgery
Arnaud de Villeneuve Hospital
Montpellier, France
Chapters 6, 12

Marzia Leache, M.D.
University of Rome "La Sapienza"
Policlinico Umberto I, Hospital
Rome, Italy
Chapters 3, 10

Patrick Limoges, M.D.
Assistant Professor of Anaesthesiology
University of Montreal
Department of Anaesthesiology
Sacré-coeur Hospital
Montreal, Québec, Canada
Chapters 5, 9

Louis P. Perrault, M.D., Ph.D.
Assistant Professor of Surgery
University of Montreal
Department of Cardiac Surgery
Montreal Heart Institute
Montreal, Québec, Canada
Chapter 7

Peter Sheridan, M.D.
Associate Professor of Surgery
University of Montreal
Department of Anaestheiology
Montreal Heart Institute
Montreal, Québec, Canada
Chapters 5, 9

Kenton J. Zehr, M.D.
Associate Professor of Surgery
Mayo Clinic College of Medicine
Division of Cardiac Surgery
Mayo Clinic
Rochester, Minnesota, U.S.A.
Chapter 13

PREFACE

The recent revival of the coronary artery bypass technique without extracorporeal assistance constitutes one of the major revolutions in the cardiovascular community in the last decade. Criticized by some[1] and praised by others,[2] off-pump surgery has left nobody indifferent. By simply typing "off-pump surgery" on a current electronic research database, close to 1,000 manuscripts can be captured indicating the gigantic infatuation with this procedure. From the first aorto-to-coronary bypass performed on a dog by Alexis Carrel in 1910 to the current technique of coronary revascularization, a lot has been accomplished and, in the future, a lot more will be.[3]

As surgeons, new techniques could be seen as a transgression of our secure and highly predictable daily environment. On the other hand, changes are possible only if we are responsive to our environment. The human being is genetically modified to be resistant to change. We just have to remember that it took us a million years to evolve from Homo Habilis to Homo Erectus and this one, who, because he had a bigger brain of 900 cc (apparently ours is 1400 cc) and longer legs, is believed to have walked his way out of Africa and initiated peopling of the world. However, to accomplish this last task, it took him another million years. On the other hand, Homo Sapiens Neanderthals had a brain comparable to ours (even bigger), existed for 240,000 years (we have been around for less than 150,000 years), but never made it to the moon. We are capable of incredible adaptation. Hundred thousand years ago we elaborated our evolutionary strategy by opting for a pair of "pliers" and two frontal lobes. In those early days, with the constant threat of a very hostile environment, the long-term success of such a strategy was not obvious. Standard and Poors would not have given us an enviable quotation on Wall Street! We were poorly equipped to rule our world then, but we had great abstraction capacity and an extraordinary ability to create and construct.

Surgeons have always been on the edge of new techniques and technologies that could facilitate, secure, and improve their "environment". This perpetual search for a "better way" has led them to relentlessly re-question and re-challenge their thinking. OPCAB surgery is another link in the long chain of progress that has characterized the history of surgery during the last century. The future will tell us if this was a good "strategy" or not. In the meantime, it is our responsibility to challenge the technique, optimize it, test its limits, and ultimately find its place in the vast puzzle of our surgical evolution.

Raymond Cartier, M.D.
Montreal Heart Institute

References

1. Jegaden O, Mikaeloff P. Off-pump coronary artery bypass surgery. The beginning of the end? Eur J Cardio-Thorac Surg 2001; 19(3):237-8.
2. Buffolo E. Why is "off-pump" coronary artery bypass grafting better? Heart Surg Forum 2002; 5(2):154-6.
3. Carrel A. On the experimental surgery of the thoracic aorta and the heart. Ann Surg 1910; 52:83-95.

FOREWORD

"Le retour aux sources"

After having been in surgical practice for more than 45 years, this new popularity of off-pump coronary artery surgery brings back many memories. In the old days, it was quite popular to operate on a "beating heart" since no other choice was offered to us. As a resident training in cardiothoracic surgery under Dr. Claude S. Beck at Case Western University (Cleveland, Ohio) in 1951, I had the opportunity to witness some of the first attempts to revascularize the myocardium without assistance of the extracorporeal circulation. During those days, Beck was working on the clinical application of some surgical procedures he had developed in the lab. The "Beck I" operation or "poudrage" consisted of pouring abrasive compounds, such as bone powder, on the surface of the epicardium to promote inflammation and neovascular interconnections between the epicardial coronaries. Beck used to say that coronary insufficiency was a disease of the "surface" vessels mainly affecting the proximal portions of the epicardial coronary arteries that run on the external part of the myocardium. By promoting neovascularization at the epicardium level, one could expect to improve blood flow beyond the obstructive lesions. But the most spectacular surgical procedure I observed him perform was this formidable operation consisting of direct anastomosis (with no interposition graft) between the coronary sinus and the descending thoracic aorta. This was called the Beck II operation. With a specially-designed, double-articulated, side-bite clamp, he would progressively approximate the distal portion of the coronary sinus to the descending thoracic aorta. Then anastomosis was performed with interrupted 6-0 silk sutures. This was done through the left chest and, without a doubt, was the most difficult anastomosis I witnessed. The purpose of the operation was to improve blood flow to the myocardium, using the venous network of the heart that was normally free of disease. Unfortunately, the clinical outcome of these operations did not fulfill Beck's expectations and were abandoned. Interestingly, as a fellow in Beck's service and taking calls, I had to do elective admissions in the surgical ward. Some patients were admitted to have previously-constructed carotido-jugular fistula taken down. Beck, being trained as a neurosurgeon, was interested in the treatment of mentally-challenged patients. By surgically creating an arterio-venous fistula between the jugular vein and the carotid artery, he expected, as in the myocardium, to improve brain oxygenation and the mental status of these patients. Unfortunately, this procedure was found to be inefficient and rapidly abandoned by Beck.

Few years later, through a common friend, I was introduced to Dr. Thomas J.E. Oneill, a colleague of Dr. Charles P. Bailey, from Philadelphia. Bailey and his team were pioneers in mitral valve surgery. They had developed a right chest approach for close heart commissurotomy in cases where the left atrial appendix approach was contraindicated.[1] I visited him to familiarize myself with the technique. It was during this visit that he informed me about a "new operation" developed and popularized by the Italian surgeons Battezzati, Tagliaferro, and De Marchi[2] to treat coronary insufficiency. Their quite simple procedure consisted of bilateral ligature of the internal thoracic arteries (ITA) for the purpose of improving the coronary artery collateral circulation. The surgery was appealing because it was technically easy, could be done on local anesthesia, and did not require intensive care unit stay. We were naive during these days! Although our clinical impression was that the patient's angina improved, Dimond put an end to our innocent hopes. In a well-designed study, he and his group from Kansas City showed that ITA ligature was as efficient as a superficial incision on the patient's skin.[3] We had just discovered the virtue of the placebo effect!

Later, as an appointed thoracic surgeon at Hotel-Dieu Hospital in Montreal, Canada, I introduced in my practice the Vineberg procedure for treating angina pectoris. Montreal was the hometown of Arthur Vineberg, the surgeon who developed and mastered this operation while working then at the Montreal Heart Institute and the Royal Victoria Hospital. Although the Vineberg procedure took time to be recognized, it acquired considerable credibility when a report from the Cleveland Clinic angiographically confirmed the patency of the ITA graft. In the early sixties, coronary angiography was in its early beginnings and our hospital barely had the knowledge and expertise in using it. Nevertheless, Dr. J.D. Mignault, one of the cardiologists working with us, had been able to perform a selective angiogram of the ITA graft in a patient showing a broad connection between the ITA graft and an obtuse marginal branch, replicating the Cleveland experience. This encouraged us to pursue the Vineberg procedure, which remained popular in Montreal until the early seventies, when direct coronary artery revascularization under extracorporeal circulation was definitely adopted. Always in search of a direct way of revascularizing the myocardium, in 1958, I attempted an endarterectomy of the left anterior descending artery on the beating heart. For several reasons, this was a difficult operation, and although the patient survived the procedure, it was not successful in re-establishing blood flow through the anterior wall. Unfortunately, coronary stabilizers were not fashionable at that time.

It is difficult to predict what would have happened if we had been ingenious enough to create stabilizers. Would that have stopped the progression and improvement of the extracorporeal circulation? It is likely though that this technology could have been available even back in those dark ages, since the underlying principle is pure mechanics. However, we were not familiar with coronary anastomosis, polypropylene threads were not available, and needle holder design was closer to a Neanderthal club than the fine and delicate microsurgical needle holders used today. It is obvious that the present generation of cardiac surgeons has a distinct advantage over us. They were already trained to perform a quick, technically-excellent anastomosis on the still heart before attempting revascularization on the beating heart.

It gives me pleasure to know that surgeons are always looking for improvement and are always willing to explore new avenues and new means to improve the quality of their work. Having myself witnessed half a century of innovations in the cardiovascular field, this "retour aux sources" brought by off-pump coronary artery surgery keeps me brimming with confidence that the new generation of surgeons will carry on further the burning flame of innovation and pursue the work of their predecessors.

Paul Cartier, M.D.
Department of Cardiovascular Surgery
Centre Hospitalier de l'Université de Montreal (CHUM)
Hotel-Dieu Hospital
Montreal, Quebec, Canada

References

1. Bailey CP, Bolton HE, Morse DP. The right approach to the problem of mitral stenosis. Surg Clin N Am 1956; 103:931-54.
2. Battezzati M, Tagliaferro A, De Marchi G. The ligature of the two internal mammary arteries in disorders of vascularization of the myocardium. Minerva Med 1955; 46(Part 2):1173-80.
3. Dimond EG, Kittle CF, Crockett JE. Comparison of internal artery ligation and sham operation for angina pectoris. Am J Cardiol 1960; 5:483-6.

CHAPTER 1

Historical Considerations

Raymond Cartier

Recently, performing coronary artery surgery on the beating heart received prime attention even though the concept is evidently not a new one. In 1910, Alexis Carrel (Fig. 1) was the first to propose bypass surgery to correct angina pectoris,[1] "In certain cases of angina pectoris, when the mouth of the coronary arteries is calcified, it would be useful to establish a complementary circulation for the lower part of the arteries". Carrel experimentally put forth effort to develop coronary artery bypass surgery on the beating heart of a dog: "I attempted to perform an indirect anastomosis between the descending aorta and left coronary artery. It was, for many reasons, a difficult operation". It took him 5 minutes to complete the distal anastomosis but the heart started fibrillating after 3 minutes of ischemia. This was due to cross clamping the entire pedicle of the heart to obtain a bloodless field. Nevertheless, he succeeded by massaging the heart to keep the dog alive for 2 hours. Later, Zoll[2] confirmed Carrel's premise by showing that in about 50% of fatal coronary artery occlusion cases, the occlusion had occurred in the proximal part of the left coronary artery network, leaving a distal coronary system suitable for imaginative surgical reconstruction. This would contribute to launching the modern era of direct coronary revascularization surgery. However, even before that time, others had already investigated indirect means of supplementing the deficient coronary artery system.

In 1932, Claude S. Beck[3] experimented on the production of inflammation on the surface of the heart for the purpose of creating anastomosis between epicardial coronary arteries. These intercoronary communications were created by abrasion and were believed to bring a more "equitable distribution of the blood" to the myocardium. To abrade the surface of the heart and remove the epicardium, Beck applied powdered beef bone. He claimed success in an animal model of coronary ischemia when a better survival rate along with infarct size reduction was observed in animal hearts that had been "prepared" prior to coronary ligation. Soon, he brought these techniques to clinical setup and reported to have used them in 37 patients afflicted with severe coronary artery disease.[4,5] During the same period, O'Shaughnessy[6] experimented on a similar indirect mean of improving the coronary circulation by attaching to the heart a pedicled graft of the great omentum, an intervention that he called: "cardio-omentopexy". Out of 15 patients treated for angina, O'Shaughnessy reported long-term disease-free survival in 8 patients. In the forties, Mercier Fauteux,[7] then working at McGill University in Montreal and inspired by the efforts of Gross[8] and Oppel,[9] experimented on ligation of the "vena magna cordis*" of the heart to improve collateral circulation. In 1913, Oppel reported successful treatment of ischemic gangrene of the foot by ligating the popliteal vein in 6 patients affected by this disease. For a while, ligature of the artery and companion vein became a fundamental principle in vascular surgery because it was shown, experimentally, to increase residual arterial pressure, venous pressure and flow in the peripheral arterial bed.[10] Gross was a strong proponent of

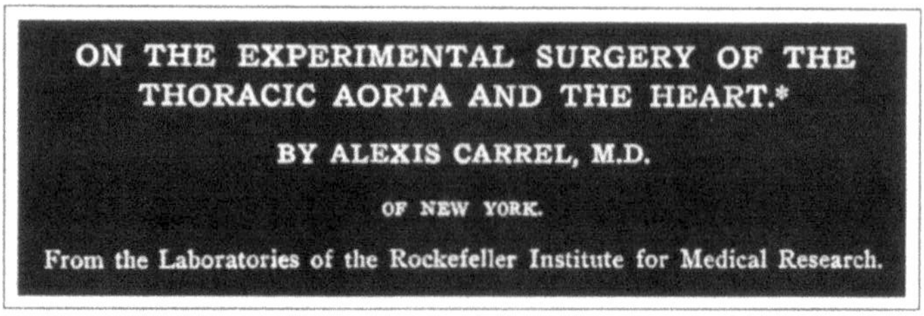

ON THE EXPERIMENTAL SURGERY OF THE
THORACIC AORTA AND THE HEART.*

BY ALEXIS CARREL, M.D.

OF NEW YORK.

From the Laboratories of the Rockefeller Institute for Medical Research.

Figure 1. Alexis Carrel, Annals of Thoracic Surgery 1910. Reprinted with permission from the Society of Thoracic Surgeons. Ann Thorac Surg 1910; 52:83-95.

ligature of the coronary sinus to improve global myocardial perfusion. His initial experiments, however, were overshadowed by devastating operative animal mortality. Acknowledging the experience of these two authors, Fauteux designed a new procedure that involved ligaturing only the vena magna cordis* to improve circulation in the anterior wall of the heart without causing death of the animal. In certain cases, he combined the coronary artery ligature procedure with a pericoronary neurectomy to perform a "local" sympathectomy, thereby promoting local arterial vasodilatation. He showed that better survival could be achieved with this surgical strategy in a canine model of myocardial ischemia. Fauteux reported on 16 patients on whom he performed this technique, claiming long-term disease-free survival in a few of them.[11]

A few years later, Beck explored new means of improving collateral circulation to the myocardium. His was inspired by the experimental work of Roberts[12] on in situ additional arterial blood supply to the myocardium through the coronary sinus. Roberts connected a glass cannula proximally to either the brachiocephalic, subclavian, or innominate artery, and distally to the coronary sinus. With this technique, he reported animal survival even after acute ligature of both the left anterior descending and the right coronary arteries. Beck[13] developed a surgical procedure where a segment of vein or artery (generally the carotid artery) was sutured between the coronary sinus and the descending aorta (Fig. 2). The procedure was generally combined with a delayed coronary sinus ligature. Although patency of the vascular graft in this procedure was initially extremely arduous to maintain, Beck finally mastered the technique after a tremendous amount of experimental work. He eventually extended its use to the diseased human heart.

In the mid-forties, Vineberg,[14] then thoracic surgeon at the Royal Victoria Hospital in Montreal, designed a new way of improving myocardial blood flow. He described a method where the internal thoracic artery was dissected and transected before its free end was transplanted directly into the left ventricular wall. Vineberg's experiments demonstrated that internal thoracic artery implants could remain patent and even promote collateral circulation with native coronary vessel use. The technique was adopted extensively for humans affected by chronic coronary artery insufficiency. The long-term angiographic patency of these implantations with extensive intramyocardial arteriolar connections was further demonstrated up to 13 years after the initial operation[15] (Fig. 3). Vineberg, at that time, claimed that the operation contributed to increased life expectancy in some patients. This was perhaps the first systematic attempt to perform direct arterial revascularization on the myocardium and was the precursor of modern coronary artery revascularization.

The true era of direct coronary artery revascularization was launched with the work of Vladimir P. Demikhov[16] (Fig. 4). Demikhov's main interest was heart and lung transplantation. As

---

* The great cardiac vein that collects the blood from the anterior surface of the ventricles follows the anterior longitudinal wall, and empties in the coronary sinus.

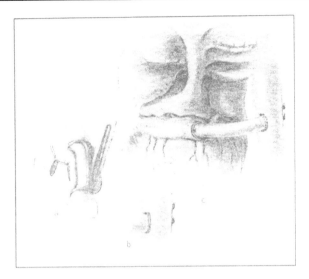

Figure 2. Beck's arterialization of the coronary sinus. Reprinted with permission from the American Medical Association. JAMA 1948; 137:436-442.

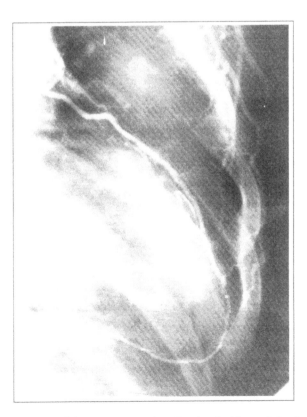

Figure 3. Vineberg's free internal thoracic artery implant. Reprinted with permission from the American Association for Thoracic Surgery. J Thorac and Cardiovasc Surgery 1975; 70:381-397.

early as 1946, he successfully achieved a canine **heterotopic** heart-lung transplantation. The secret of his success was apparently due to a very performant heart-lung preparation that was kept viable by its own closed-circuit circulation. This allowed the transplanted organ to remain alive with minimal damage during recipient preparation. In the same year, he performed the first **orthotopic** heart and lung transplantation; and in the 1951, he performed the first **orthotopic** heart transplantation. Demikhov accomplished all these world premières in a self-made laboratory without access to hypothermia or a pump-oxygenator. It was his vast experience in experimental organ transplantation that brought him to the field of coronary surgery. On April 1952, he attempted the first internal thoracic artery to coronary artery bypass on a dog heart. The experiment failed due to too long ischemic time (>2 minutes) that led to heart fibrillation and death of the animal. However, the following year, he succeeded by modifying a late 19th century technique for vascular anastomosis by Payr.[17] Payr's technique (Fig. 5) consisted of a rigid metal ring inside of which the proximal vessel was introduced and folded back before being put into the distal vessel. It could be performed very quickly, in less than a minute; however, it only permitted end-to-end anastomoses. With this technique, Demikov was able to demonstrate anastomotic patency up to 2 years after the surgery. During the same period, some thousand miles away in the New World, Gordon Murray, a Canadian surgeon, also experimented on arterial revascularization of the myocardium.

Figure 4. Vladimir P. Demikhov, a Russian surgeon pioneered in heart-lung transplantation and coronary artery surgery. Reprinted with permission from the Society of Thoracic Surgery. Ann Thorac Surg 1998; 65:1171-7.

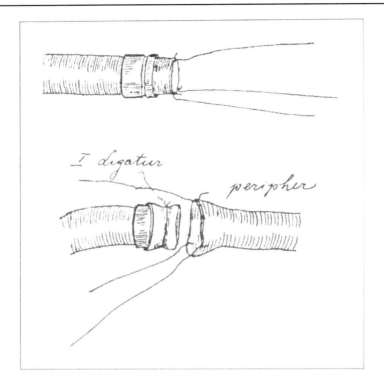

Figure 5. Payr's technique on end-to-end vascular anastomosis.

Murray imagined a perfusion system for supplying distal blood flow during coronary artery grafting. Oxygenated heparinized blood was injected via an 18-gauge needle distal to the arteriotomy during suturing of the graft.[18] Later, he managed to use an endoluminal polythene tube to maintain distal perfusion during coronary grafting. In animals, he successfully harnessed the subclavian, axillary, internal thoracic, or carotid artery as conduits to bypass the proximal coronary artery network and established arterial connection on the distal coronary bed. With this technique, Murray claimed that he could restore the coronary circulation in <30 seconds, and performed coronary artery bypass without perioperative myocardial infarction (Fig. 6).

The first clinical attempt to perform direct coronary artery revascularization probably belongs to Longmire, as reported by Westaby.[19] "At the time, we were doing the coronary thromboendarterectomy procedure, we also, I think, performed a couple of the earliest internal mammary to coronary anastomoses. We were forced into it when the coronary artery we were endarterectomizing disintegrated, and in desperation, we anastomosed the internal mammary to the distal end of the right coronary and later decided it was a good operation". Longmire did not comment on the long-term follow-up of these life-saving operations that he performed in the mid-fifties, but the course of history clearly confirmed his primary intuition. In the early sixties, Goetz performed with a modified Payr technique the first internal thoracic-to-coronary artery bypass graft on human.[20] Isolated cases of coronary artery bypass with the saphenous vein as conduit were also reported. In 1962, Sabiston reported the first successful aorta to right coronary artery bypass without extracorporeal circulation.[21] Unfortunately, the patient died a few days later of a massive stroke (Fig. 7). The following year, DeBakey and his group performed the first saphenous vein bypass on the left anterior descending artery.[22] They reported the case 7 years later with a coronary angiography confirming the patency of the graft (Fig. 8).

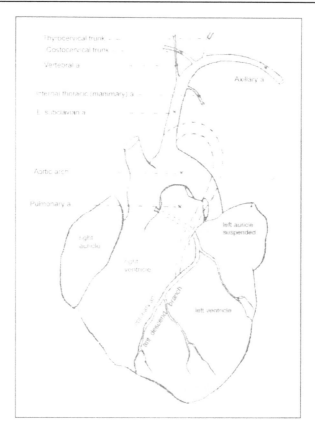

Figure 6. Gordon Murray's experimental work on coronary artery grafting. See text for explanation. Reprinted with permission from the Canadian Medical Association. Can Med Assoc J 1954; 71:594-597.

But it was Vasilii I. Kolesov (1904-1992), from the first Leningrad Medical Institute of I.P. Pavlov, who initiated, in the sixties, the modern era of coronary artery bypass grafting on the beating heart[23] (Fig. 9). Between 1964 to 1976, Kolesov and his associates performed more than 130 coronary revascularizations, most of them without extracorporeal circulation. Kolesov was the first to confirm that direct coronary grafting with internal thoracic arteries (ITA) (single and bilateral) could treat stable and unstable angina as well as evolving myocardial infarction. With V. F. Gudov, he designed a circular vessel suturing apparatus for mechanical suturing of ITA conduits. They later improved this apparatus, and it became known as the VCA-4 model (Fig. 10). Its improvement enabled them to create end-to-end anastomosis between ITA and coronary arteries from 1.3 to 4 mm in diameter in an average of 5 to 7 minutes. The most amazing element of Kolesov's experiments was that he performed all these operations without preoperative coronary angiography. Target vessels were picked according to EKG findings and surgical digital exploration. Coronary disease was detected through palpation and visualization of "bead like nodules" and "dimpling" of the epicardium in the neighborhood of the coronary vessels.

In May 1967, Rene G. Favaloro initiated the modern era of coronary artery revascularization by adapting the extracorporeal circulation to support patient hemodynamics

Figure 7. Sabiston's first case of aorto-coronary grafting to the right coronary artery network. Reprinted with permission from John Hopkins Medecine Publications. John Hopkins Med J 1974; 134:314.

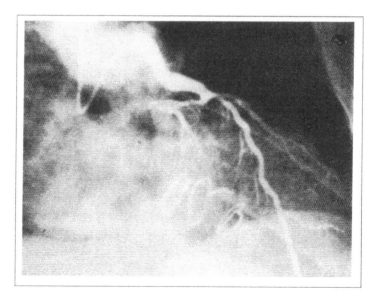

Figure 8. Coronary angiogram showing patency of Garret's first case of aorto-coronary artery bypass to the left anterior descending artery. Reprinted with permission from the American Medical Association. JAMA 1973; 223:792-94.

Figure 9. Russian surgeon Vasilii I. Kolesof whose extensive clinical experience in off pump bypasses surgery set off the modern era of coronary artery revascularization. Reprinted with permission from the Association for Thoracic Surgery. J Thorac and Cardiovasc Surgery 1967; 54:535-544.

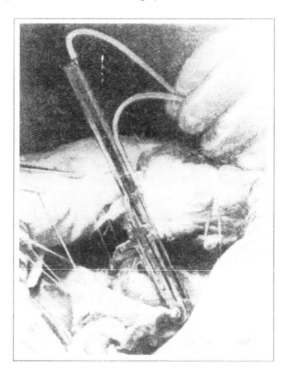

Figure 10. Kolesof's vessel-suturing circular apparatus performing end to end anastomosis. Reprinted with permission from the Association for Thoracic Surgery. J Thorac and Cardiovasc Surgery 1967; 54:535-544.

during graft suturing.[24] In 1968, he reported on 155 patients in whom 222 coronary artery bypasses were carried out employing either saphenous vein grafts or internal mammary artery implants with an operative mortality of 5%. However, others pursued in performing off-pump bypass in single and double diseases mainly to avoid inflammatory side-effects during those early days of cardiopulmonary bypass. In 1975, Trapp and Bisarya from Vancouver, B.C., to avoid cardiopulmonary bypass-related side-effects, proposed an original technique for off-pump grafting deploying internal mammary artery and saphenous vein conduits.[25] To avoid hemodynamic unstability related to regional ischemia during grafting, they placed a perfusing cannula in the distal anastomotic site (Fig. 11). The cannula was connected proximally to the ascending aorta, and could be used distally for either antegrade or retrograde perfusion of the targeted vessel. They reported on 63 patients in whom no cardiopulmonary bypass assistance was sought, with only one death due to a cerebrovascular accident.

In the seventies and eighties, several groups maintained their commitment to off-pump coronary surgery mainly for economical purposes.[26-28] Most of these series reported on selected cases, generally patients with single and double vessel disease excluding the posterior territory. Surgeons had to wait for the commercialization of coronary stabilizers to

Figure 11. Trapp's perfusion system during off pump coronary artery. Reprinted with permission of the Society of Thoracic Surgery. Ann Thorac Surg 1975; 19:1-9.

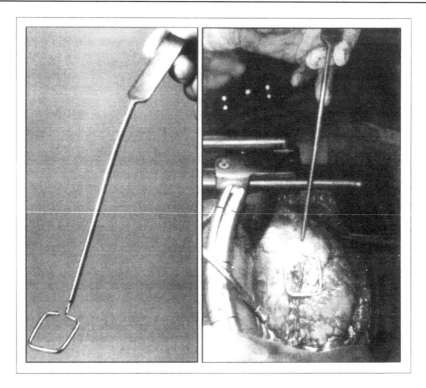

Figure 12. Grondin's coronary artery stabilizer. Reprinted with permission of Seminars in Cardiothoracic vascular Anesthesia. Sem Cardiothorac Vascular Anesthesia 2000; 4:103-109.

safely access the posterior wall and enable complete revascularization for patients with multivessel disease. Interestingly, before the intensive commercialization of "coronary stabilizers" of the mid-nineties, some authors had already reported the development of "home-made" stabilizers to aid in coronary artery surgery (Fig. 12). In 1975, Pierre Grondin from the Montreal Heart Institute developed his own version of a coronary stabilizer to control back bleeding during coronary anastomosis.[29] Others presented similar stabilizing devices, which to some extent, strangely resemble those being used in our contemporary era (Figs. 13,14).[30,31] For instance, Dr. Fayes Aboudjaouday from Libanon developed in 1992 coronary artery stabilizers for both anterior and lateral territory as well as an apical suction cup that were quite in advance for the time (Fig. 15).[32] In this current era, numerous commercialized retractor stabilizers have been refined and proven to be efficient in providing surgeons with the ability to achieve complete revascularization, even in patients with circumflex artery disease.

Today, off-pump coronary artery surgery is part of the armamentarium of modern cardiac surgery. Although its revival was not without generating controversies and debates in the cardiovascular community, it has contributed to improve our knowledge and strategies in the treatment of coronary insufficiency.

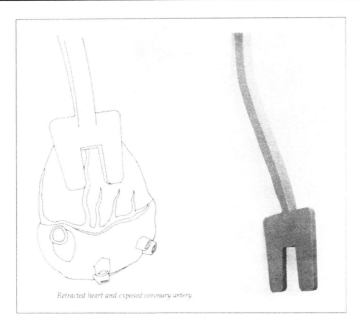

Figure 13. Delrossi and Lemolle myocardial retractor. Reprinted with permission from the Society of Thoracic Surgery. Ann Thorac Surg 1983; 36:101-2.

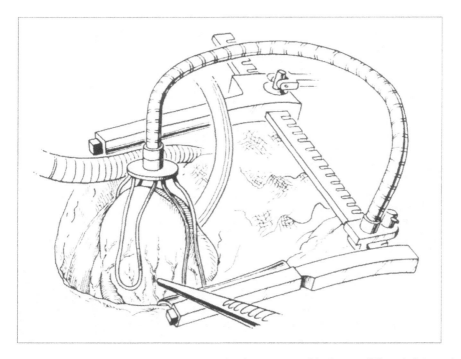

Figure 14. Roux's myocardial "holder". Reprinted with permission of the Society of Thoracic Surgery. Ann Thorac Surg 1989; 48:595-6.

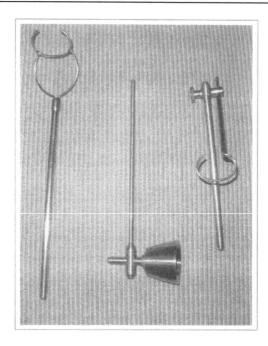

Figure 15. Dr. Fayes Aboudjaouday's instruments developed in 1992.

References

1. Carrel A. On the experimental surgery of the thoracic aorta and the heart. Ann Surg 1910; 52:83-95.
2. Zoll PM, Wessler S, Schlesinger MJ. Interarterial coronary anastomoses in the human heart, with particular reference to anemia and reıative cardiac anoxia. Circulation 1951; 4:797-812.
3. Beck CS, Tichy VR. Production of collateral circulation to the heart: Experimental study. Am Heart J 1935; 10:849-73.
4. Beck CS. Communications between the coronary arteries produced by application of inflammatory agents to the surface of the hearts. Ann Surg 1943; 118:34-45.
5. Feil H. Clinical appraisal of the Beck operation. Ann Surg 1943; 118:807-15.
6. O'Shaughnessy L. An experimental method of providing a collateral circulation to the heart. Heart Br J Surg 1936; 23:665-70
7. Fauteux M. Experimental study of the surgical treatment of coronary disease. Surg Gynecol Obstet 1940; 71:151-60
8. Gross L, Blum L, Siverman G. Experimental attempts to increase the blood supply to the dog's heart by means of coronary sinus occlusion. J Exp Med 1937; 65:91-110.
9. Oppel WA. Wieting's Operation und der reduzierte Blutkreislauf. Wratschebnaja Gaz 1913; 40:1241.
10. Makins G. Hunterian operation. Lancet 1917; 1:249-254
11. Fauteux M. Surgical treatment of angina pectoris. Experiences with ligation of the great cardiac vein and pericoronary neurectomy. Ann Surg 1946; 124:1041-6
12. Roberts JT, Browne RS, Roberts G. Nourishment of the myocardium by way of the coronary veins. Fed Proc 1943; 2:90.
13. Beck CS, Stanton E, Batiuchok W et al. Revascularization of heart by graft of systemic artery into coronary sinus. JAMA 1948; 137:436-42.
14. Vineberg AM. Development of an anastomosis between the coronary vessels and the transplanted internal mammary artery. Can Med Assoc J 1946; 55:117-9.
15. Vineberg A. Evidence that revascularization by ventricular-internal mammary artery implants increases longevity, twenty-four year, nine-month follow-up. J Thorac Cardiovasc Surg 1975; 70:381-97.

16. Konstatinov IE. A mystery of Valdimir P. Demikov: The 50th anniversary of the first intrathoracic transplantation. Ann Thorac Surg 1998; 65:1171-7.

17. Payr E. Beiträge zur technik der blutgefäss- und nervennaht nebst mittheilungen über die verwendung eines resorbirbaren metalles in der chirurgie. Archiv für Klinische Chirurgie 1901; LXII:S67-S93.

18. Murray G, Porcheron R, Hilario J et al. Anastomosis of a systemic artery to the coronary. Can Med Assoc J 1954; 71:594-7.

19. Westaby S. Coronary surgery without cardiopulmonary bypass. Br Heart J 1995; 73:203-4.

20. Goetz RH, Rohman M, Haller JD et al. Internal mammary-coronary artery anastomosis. A nonsuture method employing tantalum ring. J Thorac Cardiovasc Surg 1961; 41:378-86.

21. Sabiston DC. The coronary circulation. John Hopkins Med J1974; 134:314-29.

22. Garret HE, Dennis EW, DeBakey ME. Aortocoronary bypass with saphenous vein. JAMA 1973; 223:792-4.

23. Olearchyk AS, Kolesov VI. A pioneer of coronary revascularization by internal mammary-coronary artery grafting. J Thorac Cardiovasc Surg 1988; 96:13-8.

24. Favaloro RG. Saphenous vein graft in the surgical treatment of coronary artery disease. J Thorac Cardiovasc Surg 1969; 58:178-85.

25. Trapp WG, Bisarya R. Placement of coronary artery bypass graft without pump oxygenator. Ann Thorac Surg 1975; 19:1-9.

26. Ankeney JL. To use or not to use the pump oxygenator in coronary artery bypass operations. Ann Thorac Surg 1975; 19:108-9.

27. Benetti FJ, Naseli G, Wood M et al. Direct myocardial revascularization without extracorporeal circulation: Experience in 700 patients. Chest 1991; 100:310-6.

28. Bufalo E, Andrade JCS, Branco JNR et al. Coronary artery bypass grafting without cardiopulmonary bypass. Ann Thorac Surg 1996; 61:63-6.

29. Cartier R. From idea to OR: Surgical innovation, clinical application and outcome. Sem Cardiothorac Vascular Anesthesia 2000; 4:103-9.

30. Delrossi AD, Lemole GM. A new retractor to aid in coronary artery surgery. Ann Thorac Surg 1983; 36:101-2.

31. Roux A, Fournial G, Glock Y et al. New helper instrument in cardiac surgery. Ann Thorac Surg 1989; 48:595-6.

32. Personal communication.

CHAPTER 2

Cardiopulmonary Bypass and Inflammation

François Dagenais

I n the 1920s, Brukhonenko was the first to advance the concept of total body perfusion with removal of the heart.[1] However, the development of a true heart-lung machine could not be fully explored until the emergence of new scientific knowledge, such as blood compatibility, identification of a reliable anticoagulant (heparin) and anticoagulant antagonist (protamine), the establishment of metabolic needs during hypothermia and implementation of the roller pump.[2] Inspired by the tragic death of a young gravid woman sustaining a massive pulmonary embolus, Dr. John Gibbon Jr was the first to suggest coupling the extracorporeal circulation with an oxygenator to conduct cardiac surgery procedures.[3] Through constant refinement of his heart-lung machine model over almost two decades, Dr. Gibbon successfully performed an atrial septal defect closure in a young woman in May 1953, using total extracorporeal circulation.[4] Such a pioneering contributive association with the works of others, such as Lillehei on the cross-circulation, Bigelow on deep hypothermia and Melrose on myocardial protection, prompted a new era in the surgical management of cardiac disorders.[5-7]

Since its clinical application, cardiopulmonary bypass (CPB) has evolved tremendously through a collaborative effort of physiologists, pharmacologists, engineers and physicians. The bulky, cumbersome, "stationary film oxygenator" of the Gibbon-Mayo-IBM pump was abandoned for the more sophisticated and efficient bubble oxygenators. The roller pump coupled with large-bore flexible tubings established itself as the optimal blood-pumping system. The explosion of coronary artery bypass surgery in the 1970s led to the industrial production of membrane oxygenators through major advances in material science and device technology. During the 1980s, recognition of the deleterious effects of CPB focused efforts on developing more biocompatible surfaces, such as heparin coating of the extracorporeal circuit surfaces.[8] Within three decades, the heart-lung machine transformed itself from an artisanal hand-made unit to an effective, reliable, large-scale mass-production system, allowing complex and lengthy cardiac surgical procedures to be undertaken.

Pathophysiology of Cardiopulmonary Bypass

Refinements of CPB in the past 30 years have emboldened surgeons to undertake the repair of complex cardiac defects. The motionless and bloodless operative field provided by CPB permits almost unlimited access to intracardiac and extracardiac structures. However, exposing blood elements to foreign surfaces of the extracorporeal circulation induces a sequence of deleterious reactions, such as the activation of plasma enzyme systems, the generation of vasoactive substances, protein denaturation and the introduction of foreign materials.[9-11] The inflammatory response triggered by CPB provokes a series of hematological and immunological cascades as well as induces post-reperfusion organ dysfunction mainly within the lungs,[12] kidneys,[13] and the neurological system.[14] In this chapter, we will initially discuss

Off Pump Coronary Artery Bypass Surgery, edited by Raymond Cartier. ©2005 Eurekah.com.

the relevant alterations in hematological and immunological functions during CPB, and subsequently address the impact of the "whole-body inflammatory response" on different organ functions.

Hemostasis

The deleterious effects of CPB on hemostasis are easily illustrated by massive clotting of the extracorporeal system if anticoagulation is omitted before CPB initiation. On the other hand, anticoagulation with heparin, coupled with rigorous operative anticoagulation monitoring, allows CPB to be performed with minimal side-effects in most circumstances. Despite the advances in extracorporeal circulation, a significant proportion of patients suffer from bleeding-related issues, such as reexploration for hemostasis and transfusion of homologous blood. The nonendothelialized surface of the extracorporeal system is largely responsable for the hemostatic defects encountered during CPB. The extracorporeal surface activates platelets and initiates a sequence of serine proteases comprising the coagulation fibrinolytic cascade. Through activation of the factor XII-high molecular weight kininogen–prekallikrein complex, the intrinsic coagulation pathway provides the most important coagulation stimulus during CPB. On the other hand, tissue factor exposed in wounds during open cardiac surgery stimulates thrombin generation through the extrinsic coagulation pathway. Activation of the extrinsic pathway is initiated by binding of tissue factor to factors VII and VIIa. Through consumption and hemodilution, coagulation factor levels tend to decline during CPB. Factor XII decreases by 70% after CPB.[15] However, through release from platelet granules or endothelial cells, factor VIII and von Willebrand factor (vWF) tend to be at or above preCPB levels.[16,17] VWF may promote hemostasis by facilitating platelet adhesion to endothelial surfaces. It has been reported that patients with vWF levels < 1.2 U/ml tend toward excessive postoperative bleeding.[17]

Fibrinolysis during CPB is initiated through contact activation of factor XII or kallikrein, cleaving plasminogen to plasmin, or by stress-induced release of tissue-type plasminogen activator. Activation of the fibrinolytic pathway during CPB is demonstrated by the progressive increase in fibrin degradation products. After CPB discontinuation, fibrinolysis rapidly subsides. Thus, persistent fibrinolysis after CPB rarely contributes to significant postoperative bleeding.[18] However, plasmin generation during CPB contributes to platelet loss and dysfunction.[19]

Although CPB activates the coagulation pathways and the fibrinolytic system, the major determinant of postoperative bleeding is linked to platelet alterations seen with CPB. Platelet adhesion and aggregation initiated by the nonendothelialized surface of the extracorporeal system, coupled with CPB-induced hemodilution, contribute significantly to the 30 to 50% platelet destruction observed during CPB.[20] In addition to the quantitative platelet decrease during CPB, qualitative platelet dysfunction is reported. Intrinsic and extrinsic factors alter the functional state of platelets during CPB. Factors responsible for platelet activation during CPB are multiple; among others, thrombin generation, activated complement, hypothermia, epinephrine and serotonin. In addition, a conformational change in platelet membranes and a temporary depletion of constitutive platelet glycoproteins are implicated in CPB-induced platelet dysfunction.[21] The end result of platelet alterations seen during CPB is an increase in bleeding time for up to 4-12 hours after cessation of CPB.[22]

Inflammatory Response Related to CPB

Cardiac surgery and CPB activate a complex network of inflammatory cascades. Mechanisms underlying the inflammatory response during CPB are complexe and still under investigation. Both humoral and cellular components of the immune system participate in activation, proliferation and propagation of the inflammatory response. Complement activation with release of C3a and C5a anaphylatoxins, plays a pivotal role in triggering the inflammatory process during CPB. C3a and C5a release associated with activation of the kallikrein and the coagulation pathways leads to activation of amplification systems with the production of proinflammatory substances, such as interleukin (IL)-1-6 and -8 and tumor necrosis factor (TNF).[23] These mediators promote increased vascular permeability, hightened adhesion of circulatory leukocytes to the vascular endothelium (through up-regulation of the adhesion molecules E selectin and ICAM-1) as well as leucocyte activation.[24] Cytokine production by monocytes and macrophages has also been linked to the identification of circulating endotoxins during CPB.[25] Splanchnic hypoperfusion during CPB may augment gut permeability and provoke bacterial translocation. Endotoxin is a potent stimulant of endothelial cell activation, resulting in up-regulation of adherence molecules and tissue factor, and causing the amplification of the cytokine release.[26] Endotoxemia could be implicated as an initiating event in the inflammatory cascade occuring during CPB.[27,28] Cremer and colleagues,[29] comparing a group of patients developing a perioperative systemic inflammatory response (SIRS) characterized by high cardiac output and low systemic vascular resistance, to a group of patients with normal hemodynamics, showed a correlation between SIRS and blood levels of IL-6. However, as opposed to other authors,[27,28] the incidence of endotoxin-positive patients was not more prominent in the SIRS group.

The impact of cardiac surgery on the immunological system is profound. Neutrophil function after CPB is disrupted. Neutrophil chemotaxis, phagocytosis and intracellular digestion have been reported to decline.[30] Decreased macrophage bacterial clearance capacity has also been noted with CPB.[31] T-cell lymphopenia, after CPB, is known to limit IL-2 production. IL-2 is important for the transformation of B lymphocytes to immunoglobulin-secreting plasma cells as well as an activator of cytotoxic T lymphocytes and natural killer cells.[32,33] Thus overall immunocompetence is attenuated after CPB, creating an imbalance between host defense and the propensity for infection.

CPB and cardiac surgery have important pathophysiological effects on the endocrine system. Surgical stress, hypothermia, hemodilution and nonpulsatile flow induce the release of stress hormones. Circulatory catecholamines have been reported to increase between two to 10-fold during CPB.[34] Hightened natriuresis observed during CPB is associated with an early rise in aldosterone secretion, followed by a late elevation of atrial natriuretic factor.[35] The rise in cortisol levels during cardiac surgery is gradual, and peaks on the first operative day.[36] An euthyroid sick syndrome state after CPB has been described. Administration of exogenous T3 after CPB has produced beneficial results, mostly in terms of hemodynamic improvement.[37] Such endocrine derangements during cardiac surgery have marked consequences on the microcirculation, clotting mechanisms, and the immune system.

Impact of the Inflammatory Response on Organ Function after CPB

As demonstrated, the inflammatory response during CPB is complex and not yet fully characterized. Substantial postoperative morbidity during cardiac operations seems related to be SIRS-related. On the other hand, assessing the "true" participation of CPB in the inflammatory response during cardiac surgery has always been difficult until the advent of off-pump revascularization. In this section, we will review the effects of CPB on organ function. Off-pump and on-pump inflammatory responses and organ dysfunctions will be compared in later chapters.

The lungs, brain and, to a lesser extent, the kidneys and gut have been identified as primary targets of the inflammatory response during CPB. Moreover, increasing data suggests that myocardial function may be impeded by cytokine release during CPB. Hennein and colleagues[38] demonstrated a 10-fold increase in circulating IL-6 in patients whose ventricular function declined after undergoing coronary artery bypass surgery. Moreover, there is evidence that TNF is synthesized within the myocardium. Through an enhancement of nitric oxide production within cardiac myocytes, TNF may provoke a negative inotropic effect and may thus be involved in the postischemic myocardial "stunning" phenomenon.[39] In addition to their deleterious actions on the myocardium, proinflammatory cytokines may alter the peripheral circulation by inducing a low peripheral resistance state which may potentially decrease perfusion to organs such as the kidneys, liver and gut.

Lung dysfunction after cardiac surgery is a significant cause of morbidity. The etiology of lung dysfunction after open heart surgery is thought to be multifactorial, occurring principally as a result of the combined effects of anesthesia, CPB and surgical trauma. The inflammatory response, generated during CPB through activation of complement and the release of oxygen-free radicals, proteases, leukotrienes and other cytokines, has been associated with widening of the alveolar-arterial oxygen gradient.[40] However, Cox and colleagues[41] comparing patients revascularized on- and off-pump, reported a similar deterioration in gas exchange postoperatively. This suggests that the pulmonary impairment observed after coronary revascularization is associated with factors other than CPB per se. Reduced chest wall and lung compliance, phrenic nerve or diaphragmatic dysfunction, pleural effusion, muscle weakness, post-transfusion or drug reactions, even cardiogenic pulmonary edema may all contribute to postoperative ventilation:perfusion mismatch. On the other hand, in rare circumstances, the inflammatory response elicited by CPB is associated with a clinical picture of adult respiratory distress syndrome (ARDS). The incidence of massive "pulmonary capillary leak" after CPB is low, probably under 2%.[42] Furthermore, the majority of patients developing ARDS after cardiac surgery have concomitant sepsis, multiple transfusions or aspiration, rendering it difficult to implicate CPB as the sole contributing factor.

Adverse neurological sequelae after cardiac operations are an important source of morbidity, prolonged hospitalization, and cost. Obvious stroke after coronary artery bypass surgery has been reported to occur in approximately 2 to 3% of patients.[43] However, subtle neuropsychological impairment after cardiac surgery has been demonstrated to affect over 50% of patients.[44] More specifically, Shaw and associates[45] prospectively examined 312 patients undergoing coronary artery bypass graft (CABG) surgery. The overall stroke rate was 1.1%. However, 61% of patients presented new neurological signs, such as primitive reflexes, scotoma and areas of hypoesthesia. In addition 79% of patients manifested a significant decline in neurocognitive performance. More concerning, Murkin et al[46] in a similar study, found persistent neurobehavioral dysfunction in 35% of patients at three years postoperatively.

Mechanisms underlying CNS injury are linked to cerebral embolism and ischemic hypoperfusion. Specific areas of the brain are more vulnerable to hypoperfusion. In these regions, "watershed" infarcts may occur with low perfusion pressure during CPB. The evidence of embolization in the pathogenesis of neurological injury during CPB is even more compelling. The brain is subjected to a shower of microemboli during extracorporeal circulation. Particles of air, atherosclerotic plaque, lipid globules, platelet aggregates, foreign material fragments and calcium are generated during CPB. A correlation has been established between the amount of microemboli delivered to the brain and the severity of the postoperative neurocognitive dysfunction.[47] A substantial amount of evidence implicates manipulation of the aorta as a major source of emboli during cardiac surgery with extracorporeal circulation.[48,49] Other significant clinical variables involved in the pathogenesis of cerebral embolization, as assessed by transcranial doppler, are the presence of aortic disease, old age and length of CPB. Improvements in extracorporeal circulation, such as the use of membrane oxygenators and arterial line microfiltration, have decreased the cerebral embolic load and lowered the incidence of neurocognitive dysfunction.[50]

The inflammatory response during CPB may also contribute to potentiate the deleterious effects of particulate emboli. Cerebral edema, described in patients undergoing uncomplicated CABG, may relate to the "whole-body inflammation" noted with CPB.[51] Cerebral edema coupled with other factors, such as hypoperfusion and venous hypertension, may decrease cerebral perfusion and promote neurological dysfunction in the already cerebrovascular-compromised host. It is thus apparent that CPB induces a variety of pathological processes leading to a spectrum of neurological sequelae.

The pathophysiology of gastrointestinal (GI) and renal dysfunctions after cardiac operations is multifactorial. Hypoperfusion, loss of pulsatility during CPB, toxic drug effects and the inflammatory response to CPB have been implicated in the genesis of renal and GI dysfunctions. Although up to 30% of patients sustain reversible renal impairment after CPB, only 1 to 5% require dialysis.[52] Similarly, significant GI complications are reported in less than 1% of cardiac surgery patients.[53] Moreover, low cardiac output syndrome is frequently associated with postoperative end-stage renal failure or significant GI complications. On the other hand, gastric pH measurement by tonometry during CPB demonstrates a correlation between the severity of gastric acidosis and the duration of CPB.[54] Gastric acidosis during CPB is a marker of GI hypoperfusion and is linked to the appearance of endotoxins in the circulation.[55]

Considering the negative impact of SIRS generated by the extracorporeal circulation, strong interest has emerged to prevent such a response. Different strategies have been developed to minimize cytokine release during CPB. Corticosteroid administration, although extensively studied over the years, remains controversial. Steroid administration before CPB has been reported to markedly inhibit TNF-α, IL-6 and IL-8 production and may enhance IL-10 production, a cytokine with potentially beneficial effects, by inhibiting the further release of other proinflammatory mediators.[56] On the other side, steroids may enhance the release of endotoxins, thus counterbalancing, at least in part, their positive action.[57] Increasing the biocompatibility of the extracorporeal circulation circuit has revealed a good outcome in decreasing the SIRS. Heparin-coated circuits have been demonstrated to lower IL-6 and IL-8 levels during CPB.[58] Belboul and colleagues[59] postulated that the reduced inflammatory response observed with heparin-coated circuits may be associated with less occult myocardial ischemic damage in patients undergoing open heart surgery. However, heparin-bonded circuits with reduced heparin dosage have no effect on cytokine release during CPB.[60] Recently, promising results have been obtained with different new concepts of extracorporeal circulation circuit coatings.[61,62] The addition of hemofiltration during CPB has been found to reduce IL-6 and TNF-α levels.[63] However, leucocyte filtration during the reperfusion phase of CPB for CABG significantly decreases the number of circulating leukocytes with no effects on the levels of complement-derived anaphylatoxins, IL-6 and IL-8.[64] Another strategy to attenuate the bypass-induced inflammatory response is through drug interventions. Aprotinin has been shown to be as effective as steroids in inhibiting CPB-induced TNF-α release.[65] However, combining aprotinin with heparin-bonded circuits has little influence on the levels of circulating cytokines compared to heparin-coated circuits alone.[66]

In summary, cardiac surgery using CPB induces SIRS with a negative impact on organ function. Ongoing studies comparing the inflammatory response in patients operated with and without bypass will further delineate the true participation of CPB in the inflammatory process. Further research in the field of molecular biology will possibly allow the synthesis of compounds genetically engineered to target specific receptors such as adhesion molecules to potentially block the inflammatory cascade induced by CPB.

References

1. Brukhonenko S. Circulation artificielle du sang dans l'organisme entier d'un chien avec coeur exclu. J Physiol Path Gen 1929; 27:257-272.
2. Galleatti PM, Mora CT. Cardiopulmonary bypass: The historical foundation, the future promise. In: Mora CT, ed. Cardiopulmonary Bypass: Principles and Techniques of Extracorporeal Circulation. NY: Springer-Verlag, 1995:3-18.
3. Gibbon Jr JH. Artificial maintenance of circulation during experimental occlusion of pulmonary artery.Ach Surg 1937; 34:1105-31.
4. Gibbon Jr JH, Dobell AR, Voigt GB et al. The closure of interventricular septal defects on dogs during open cardiotomy with maintenance of cardio-respiratory functions by a pump oxygenator. J Thorac Surg 1954; 28:235-40.
5. Bigelow WG, Lindsay WK, Greenwood WF. Hypothermia: Its possible role in cardiac surgery: An investigation of factors governing survival in dogs at low temperature. Ann Surg 1950; 132:849-66.
6. Lillehei CW. Controlled cross circulation for direct-vision intracardiac surgery; correction of ventricular septal defects, atrioventricularis communis and tetralogy of Fallot. Poat Grad Med 1955; 17:388-96.
7. Melrose DG, Dreyer B, Baker JBE. Elective cardiac arrest: Preliminary communication. Lancet 1955; 2:21-2.
8. Nilsson L, Storm KE, Thelin S et al. Heparin coated equipment reduces complement activation during cardiopulmonary bypass in the pig. Artif Organs 1990; 14:46-8.
9. Kirklin JK, Westaby S, Blackstone EH et al. Complement and damaging effects of cardiopulmonary bypass. J Thorac Cardiovasc Surg 1983; 86:845-57.
10. Downing SW, Edmunds Jr LH. Release of vasoactive substances during cardiopulmonary bypass. Ann Thorac Surg 1992; 54:1236-43.
11. Butler J, Chong GL, Baigrie RJ et al. Cytokine responses to cardiopulmonary bypass with membrane and bubble oxygenation. Ann Thorac Surg 1992; 53:833-8.
12. Van Belle AF, Wesseling GJ, Penn OCKM et al. Postoperative pulmonary function abnormalities after coronary artery bypass operations. Resp Med 1992; 86:195-9.
13. Hiberman M, Derby GC, Spencer RJ et al. Sequential pathophysiological changes characterizing the progression from renal dysfunction to acute failure following cardiac operation. J Thorac Cardiovasc Surg 1980; 79:838-44.
14. Roach GW, Kanchuger M, Mangano CM et al. Adverse cerebral outcomes after coronary bypass surgery. N Engl J Med 1996; 335:1857-63.
15. Bick RL, Frazier BL, Saudners CR et al. Alterations of hemostasis during cardiopulmonary bypass: The potential role of factor XII in inducing primary fibrino(geno)lysis. Blood 1984; 64:A926.
16. Bagge L, Lilienberg G, Nyström S-O et al. Coagulation, fibrinolysis and bleeding after open-heart surgery. Scand J Thorac Surg 1986; 20:151-60.
17. Weinstein M, Ware JA, Troll J et al. Changes in von Willebrand factor during cardiac surgery: Effect of desmopressin acetate. Blood 1988; 71(6):1648-55.
18. Tuman KJ, Spiess BD, McCarthy RJ et al. Comparison of viscoelastic measures of coagulation after cardiopulmonary bypass. Anesth Analg 1989; 69:69-75.
19. Khuri SF, Valeri CR, Loscalzo J et al. Heparin causes platelet dysfunction and induces fibrinolysis before cardiopulmonary bypass. Ann Thorac Surg 1995; 60:1008-14.
20. Zilla P, Fasol R, Groscurth P et al. Blood platelets in cardiopulmonary bypass operations. J Thorac Cardiovasc Surg 1989; 97:379-87.
21. George JN, Pickett EB, Saucerman S et al. Platelet surface glycoproteins. Studies on resting and activated platelets and platelet membrane microparticles in normal subjects, and observations in patients during adult respiratory distress syndrome and cardiac surgery. J Clin Invest 1986; 78:340-8.
22. Harker LA, Malpass TW, Branson HE et al. Mechanism of abnormal bleeding in patients undergoing cardiopulmonary bypass: Acquired transient platelet dysfunction associated with selective α-granule release. Blood 1980; 56:824-34.
23. Hammerschmidt DE, Stroncek DF, Bowers TK et al. Complement activation and neutropenia occurring during cardiopulmonary bypass. J Thorac Cardiovasc Surg 1981; 81:370-7.

24. Gillinov AM, Bator JM, Zehr KJ et al. Neutrophil adhesion molecule expression during cardiopulmonary bypass with bubble and membrane oxygenators. Ann Thorac Surg 1993; 56:847-53.
25. Taggart DP, Sundaram S, McCarthy C et al. Endotoxemia, complement, and white blood cell activation in cardiac surgery: A randomized trial of laxatives and pulmonary perfusion. Ann Thorac Surg 1994; 57:376-82.
26. Boyle EM, Pohlman TH, Johnson MC et al. Endothelial cell injury in cardiovascular surgery: The systemic inflammatory response. Ann Thorac Surg 1997; 63:277-84.
27. Baue AE. The role of the gut in the development of multiple organ dysfunction in cardiothoracic patients. Ann Thorac Surg 1993; 55:822-9.
28. Rocke DA, Gaffin SL, Wells MT et al. Endotoxemia associated with cardiopulmonary bypass. J Thorac Cardiovasc Surg 1987; 93:832-7.
29. Cremer J, Martin M, Redl H et al. Systemic inflammatory response syndrome after cardiac operations. Ann Thorac Surg 1996; 61:1714-20.
30. Bubeuik O, Meakins JL. Neutrophil chemotaxis in surgical patients: Effect of cardiopulmonary bypass. Surg Forum 1976; 27:267-9.
31. Bufkin BL, Gott JP, Mora CT et al. The immunologic system: Perturbations following cardiopulmonary bypass and the problem of infection in the cardiac surgery patient. In: Mora CT, ed. Cardiopulmonary bypass: Principles and Techniques of Extracorporeal Circulation. NY: Springer-Verlag, 1995:169-185..
32. Eskola J, Salo M, Viljanen MK et al. Impaired B lymphocyte function during open heart surgery. Br J Anaesth 1984; 56:333-7.
33. Brody JI, Pickering NJ, Fink GB et al. Altered lymphocyte subsets during cardiopulmonary bypass. Am J Clin Pathol 1987; 626-8.
34. Reves JG, Karp RB, Buttner EE et al. Neuronal and adrenomedullary catecholamine release in response to cardiopulmonary bypass in man. Circulation 1982; 66:49-55.
35. Schaff HV, Masburn JP, McCarthy PM et al. Natriuresis during and early after cardiopulmonary bypass: Relationship to atrial natriuretic factor, aldosterone and antidiuretic hormone. J Thorac Cardiovasc Surg 1989; 98:979-86.
36. Taylor KM, Jones JV, Walker MS et al. The cortisol response during heart-lung bypass. Circulation 1976; 54:20-5.
37. Novitzky D, Cooper DKC, Barton CI et al. Triiodothyronine as an inotropic agent after open heart surgery. J Thorac Cardiovasc Surg 1989; 98:972-7.
38. Hennein HA, Ebba H, Rodriguez JL et al. Relationship of the proinflammatory cytokines to myocardial ischemia and dysfunction after uncomplicated coronary revascularization. J Thorac Cardiovasc Surg 1994; 108:626-35.
39. Finkel MS, Oddis CV, Jacob TD et al. Negative inotropic effects of cytokines on the heart mediated by nitric oxide. Science 1992; 257:387-9.
40. Royston D. The inflammatory response and extracorporeal circulation. J Cardiothorac Vasc Anesth 1997; 11:341-54.
41. Cox CM, Ascione R, Cohen AM et al. Effect of cardiopulmonary bypass on pulmonary gas exchange: A prospective randomized study. Ann Thorac Surg 2000; 69:140-5.
42. Messent M, Sullivan K, Keogh BF et al. Adult respiratory distress syndrome following cardiopulmonary bypass: Incidence and prediction. Anesthesia 1992; 47:267-8.
43. Bojar RM, Najafi H, DeLaria GA et al. Neurological complications of coronary revascularization. Ann Thorac Surg 1983; 36:427-32.
44. Smith PCC, Treasure T, Newman SP et al. Cerebral consequences of cardiopulmonary bypass. Lancet 1986; 1:823-5.
45. Shaw PG, Bates D, Cartlidge NEF et al. Early intellectual dysfunction following coronary artery bypass surgery. Q J Med 1986; 58(225):59-68.
46. Murkin JM, Baird DL, Martzke JS et al. Long-term neurological outcome 3 years after coronary artery bypass surgery. Anesth Analg 1996; 82(Suppl):S328.
47. Stump DA, Tegeler CH, Rogers AT et al. Neuropsychological deficits are associated with the number of emboli detected during cardiac surgery. Stroke 1993; 24(3):A509.
48. Stump DA, Rogers AT, Kahn ND et al. When emboli occur during coronary artery bypass graft surgery. Anesthesiology 1993; 79(suppl 3A):A49.

49. Pugsley W, Klinger L, Paschalis C et al. Microemboli and cerebral impairment during cardiac surgery. Vasc Surg 1990; 1:34-43.

50. Pugsley W, Klinger L, Paschalis C et al. The impact of microemboli during cardiopulmonary bypass on neuropsychological functioning. Stroke 1994; 25:1393-9.

51. Harris DNF, Bailey SM, Smith PLC et al. Brain swelling in the first hour after coronary artery bypass surgery. Lancet 1993; 342:586-7.

52. Corwin HL, Sprague SM, DeLaria GA et al. Acute renal failure associated with cardiac operations. J Thorac Cardiovasc Surg 1989; 98:1107-12.

53. Simic O, Strathausen S, Hess W et al. Incidence and prognosis of abdominal complications after cardiopulmonary bypass. Cardiovasc Surg 1999; 7(4):419-24.

54. Landow L, Phillips DA, Heard SO et al. Gastric tonometry and venous oximetry in cardiac surgery. Crit Care Med 1991; 19:1226-33.

55. Anderson LW, Landow L, Baek L et al. Association between gastric intramucosal pH and splanchnic endotoxin, antibody to endotoxin, and tumor necrosis factor-α concentrations in patients undergoing cardiopulmonary bypass. Crit Care Med 1993; 21:210-7.

56. Teoh KHT, Bradley CA, Gauldie J et al. Steroid inhibition of cytokine-mediated vasodilation after warm heart surgery. Circulation 1995; 92(suppl 2):347-53.

57. Anderson LW, Baek L, Thomsen BS et al. Effect of methylprednisolone on endotoxemia and complement activation during cardiac surgery. J Cardiothorac Anesth 1989; 3:544-9.

58. Steinberg JB, Kapelanski DP, Olsen JD et al. Cytokine and complement levels in patients undergoing cardiopulmonary bypass. J Thorac Cardiovasc Surg 1993 ; 106:1008-16.

59. Belboul A, Lofgren C, Storm C et al. Heparin-coated circuits reduce occult myocardial damage during CPB: A randomized, single blind clinical trial. Eur J Cardiothorac Surg 2000; 17:580-6.

60. Francalancia NA, Aeba R, Yousem SA et al. Deleterious effects of cardiopulmonary bypass on early graft function after single lung allotransplantation: Evaluation of a heparin-coated bypass circuit. J Heart Lung Transplant 1994; 13:498-507.

61. Tevaerai HT, Mueller XM, Seigneul I et al. Trillium coating of cardiopulmonary bypass circuits improves biocompatibility. Int J Artif Organs 1999; 22:629-34.

62. Saito N, Motoyama S, Sawamoto J. Effects of new polymer-coated extracorporeal circuits on biocompatibility during cardiopulmonary bypass. Artif Organs 2000; 24:547-54.

63. Millar AB, Armstrong L, van der Linden J et al. Cytokine production and hemofiltration in children undergoing cardiopulmonary bypass. Ann Thorac Surg 1993; 56:1499-502.

64. Baksaas ST, Flom-Halvorsen HI, Ovrum E et al. Leucocyte filtration during cardiopulmonary reperfusion in coronary artery bypass surgery. Perfusion 1999; 14:107-17.

65. Hill GE, Alonso A, Spurzem JR et al. Aprotinin and methylprednisolone equally blunt cardiopulmonary bypass-induced inflammation in humans. J Thorac Cardiovasc Surg 1995; 110:1658-62.

66. Parolari A, Alamanni F, Gherli T et al. "High" dose aprotinin and heparin-coated circuits: Clinical efficacy and inflammatory response. Cardiovasc Surg 1999; 7:117-27.

CHAPTER 3

Potential Benefit of OPCAB Surgery

Marzia Leache and Raymond Cartier

Since the introduction of the extra-corporeal circulation (ECC) in the late fifties, its use has been seen as a necessary evil to perform surgery on the heart. The systemic inflammatory reaction generated under cardiopulmonary bypass (CPB) is thought to be responsible for the so-called "reperfusion syndrome" causing, to some extent, the general organ dysfunction encountered after on-pump surgery.[1] This is considered to be a major player in post-operative morbidity and prolonged post-operative length of hospital stay. Besides triggering the activation of leucocytes, complement, and the release of pro-inflammatory cytokines, ECC also affects nitric oxide metabolism, and can promote oxidative stress injury through the release of oxygen free radicals.[2-4] Avoiding ECC may decrease these effects but not eradicate them. The single act of splitting the sternum and entering the mediastinal cavity is, per se, already a major biological aggression. The purpose of this chapter is to review the potential benefit of off-pump surgery on organ dysfunction and peri-operative morbidity in regard to the surgery-related inflammatory reaction.

Inflammatory Response and Oxidative Stress

CBP classically activates 5 plasma protein systems and 5 different type of blood cells.[5] These plasma protein systems are the contact, intrinsic coagulation, extrinsic coagulation, complement, and fibrinolytic systems. The blood cells mostly affected are platelets, neutrophils, monocytes, lymphocytes, and endothelial cells. The inflammatory reaction expresses itself mainly through activation of the contact system, the complement system, neutrophils, and monocytes. The contact system will promote the release of bradykinin and thrombin by activation of the kallikrein cascade and the intrinsic coagulation system. Complement-activation will promote the release of anaphylatoxins with vasoactive properties, and ultimately their terminal components will have lysis capability. The three anaphylatoxins (C3a, C4a, and C5a) released by complement activation will transform neutrophils into killer cells. The activated neutrophils will produce several particularly cytotoxic enzymes. Among them are elastase, a powerful cytotoxic enzyme that is especially released during CPB along with oxygen free radicals, hydrogen peroxide, and hypobromous acid.[6] Activated monocytes release tissue factor, that has procoagulant activity, and numerous pro-inflammatory and anti-inflammatory cytokines, including tumour necrosis factor (TNF).[7,8] These are all major reactions and events that are expected to be attenuated during OPCAB surgery. Several studies have shown that OPCAB surgery can modulate the inflammatory reaction normally generated by use of the extracorporeal circulation. We propose a review of the major studies published on this topic.

Brasil[9] and his group from São Paulo, Brazil, in 1998, performed one of the earliest studies on the inflammatory response after off-pump surgery. They prospectively examined 20 good risk patients assigned either to off- or on-pump surgery. Preoperatively, serum TNF was undetectable in all patients from both groups. According to their results, significant plasma TNF levels were detected in 60% of the on-pump patients and in none of the off-pump patients. Peaks occurred mainly during and early on after ECC. Clinically, they observed less fever, hypotension, post-operative bleeding, use of inotropic drugs in the off-pump cohort and shorter orotracheal intubation time. They concluded that TNF expression was a marker of the whole body inflammatory response related to the ECC and a major player in the pathophysiological alterations that occurred. Although the authors claimed that both cohorts were comparable, on-pump cases had more bypasses performed (3.1 vs. 2.1) and longer surgical time (286 min vs. 195 min, p< 0.001), which could have accounted for the differences between the 2 groups. The same year Gu et al[10] published the results of a randomized study where 62 consecutive patients with single left anterior descending (LAD) artery disease were assigned to either conventional internal thoracic artery-to-LAD bypass through sternotomy or off-pump surgery through a minimally invasive direct coronary artery bypass (MIDCAB). They looked at complement activation (C3a), leukocyte activation (elastase), and platelet degradation (β-thromboglobulin release). Significant elevation of these 3 markers were recorded only in the conventional group. Clinical outcome favoured MIDCAB in terms of length of hospital stay, blood loss, transfusions, and post-operative ventilatory support. The results of this study were quite pertinent by clearly expressing how off-pump surgery through a limited thoracotomy could contribute to decrease the inflammatory reaction related to ECC. Interestingly, this investigation did not address the issue of multivessel patients, who represent most of the daily practice of cardiac surgeons.

In 2000, Matata and colleagues[11] demonstrated a significant rise in inflammation mediators, such as complement C3a, interleukin-8 (IL-8), elastase, and TNFα among patients exposed to CPB. They studied these markers on 20 prospectively randomized patients. To isolate the specific effects of CPB, all patients underwent beating heart surgery either with or without CPB assistance. Oxidative stress markers, such as lipid hydroperoxide and protein carbonyl, were also increased in on-pump cases. Clinically, they observed a decreased white cell count, post-operative fever, weight gain, and a shorter hospital stay among OPCAB patients. They concluded that off-pump surgery on the beating heart could significantly reduce oxidative stress and inflammation related to conventional surgery. Ascione et al[12] from Bristol published their experience with the assessment of inflammation in off-pump surgery. They randomized 60 low-risk (mean Parsonnet score: 5.4 and 6.5) patients with mainly double vessel disease (graft/patient: 2.2). They directed their attention to complement (C3a, C5a) and neutrophil activation (elastase, IL-8). The inflammatory markers were sampled peri-operatively for 24 hours. IL-8 and elastase serum levels were significantly elevated in on-pump patients. Contrary to previous observations, C3a and C5a concentrations increased significantly in both groups, although there was an obvious trend favouring the off-pump group. In the latter, post-operative leukocyte count was significantly lower along with the overall incidence of post-operative infection. They concluded in favour of OPCAB surgery for both clinical outcome and biochemical markers of the inflammatory response.

According to these randomized series, it is likely that nonuse of CPB decreases the global inflammatory process related to the ECC, mainly by reducing oxidative stress and by attenuating complement, leukocyte, and platelet activation. Furthermore, this was shown clinically to have a positive impact on post-operative outcome by shortening hospital stay and by curtailing post-operative bleeding as well as the overall incidence of infection. However, all these studies were conducted on selected patients characterized mainly as "good risk candidates".

Myocardial Protection

Many authors have reported the potential benefit of off-pump surgery on myocardial protection. Pfister,[13] in a case match study, demonstrated a decreased use of the intra-aortic balloon pump and a lower incidence of low cardiac output among patients operated off-pump.[13] This has also been our experience.[15] Wan et al[16] in a prospective study of 44 consecutive patients assigned to either off or on-pump surgery, found a positive correlation between the increase of IL-8 in patients operated on-pump and myocardial specific creatine kinase phosphate (CK-MB) serum level. IL-8 is a chemokine with the capability of attracting and activating neutrophils as well as T lymphocytes.[18] Experimentally and clinically, IL-8 is released after ischemia-reperfusion injury of the myocardium as well as acute myocardial infarction.[19,20] Conversely, ischemia-reperfusion injury can be prevented experimentally by the administration of anti-IL-8 antibodies,[21] suggesting that the lower IL-8 production associated with off-pump surgery could be related to a lesser degree of neutrophil activation and ischemia-reperfusion injury.

Czerny and colleagues[17] prospectively randomized 30 patients to undergo coronary artery bypass grafting (CABG) surgery either with or without ECC assistance. They found that post-operative myoglobin, CK-MB mass and troponin I release were significantly lower in the off-pump cohort confirming Wan's data.[16] Van Dijk and Angelini confirmed in randomized studies these findings.[22-23]

Pulmonary Physiology

Post-operative disturbance of pulmonary physiology has frequently been attributed to use of the ECC. However, very few studies have explored this aspect in regards to off-pump surgery. Surprisingly, these investigations did not conclusively reach the expected benefits.

Cox et al[24] in a prospective, randomized study of 52 patients, demonstrated a comparable involvement of the pulmonary physiology without consideration of the use or nonuse of the ECC. Alveolar-arterial gradients up to 6 hours after tracheal extubation were equally affected. Taggart,[25] in a nonrandomized prospective study, reported similar findings. The alveolar-arterial gradient remained abnormal up to 5 days after surgery with maximal respiratory dysfunction at 48 hours. Concurrently, they did not detect any detrimental effect on the pulmonary physiology related to the harvesting of bilateral internal thoracic arteries.

According to these studies, the pulmonary physiology disturbances observed during CABG surgery do not seem to relate to use of the CPB apparatus, but rather to the surgical approach, sternotomy. Some benefits may be seen with smaller incisions, such as with the MIDCAB approach (small left anterior thoracotomy), as reported by Lichtenberg et al.[26] They noted better post-operative preservation of the lung's vital capacity and forced expiratory volume in 1 second (FEV1) on days 3 and 5 following surgery in MIDCAB patients compared to those operated through conventional sternotomy. The post-operative pain score also favoured the former. They did not look specifically at the alveolar-arterial gradient. Ohkado et al[27] in a similar study comparing conventional on-pump surgery to OPCAB and MIDCAB obtained similar results in favouring the MIDCAB approach. In contrast, the data from 2 recent randomized trials, comparing on- and off–pump CABG and looking at clinical endpoints, reached different conclusions. Van Djik et al[22] and Angelini et al[23] found a shorter intubation time along with a shorter intensive care unit (ICU) stay among patients operated off-pump, even though both cohorts were comparable for preoperative risk factors.

According to these studies, it appears that the surgical approach as well as the use of CPB affect post-operative pulmonary physiology. The fact, for instance, that IL-6 release is elevated equally in both approaches (on- and off-pump) when sternotomy[11,12] is used suggests that opening the chest is by itself a major insult. The development of less invasive surgical apertures will certainly be justified in the future to help preserve the pulmonary capacity of patients.

Brain Physiology

Stroke occurrence during coronary artery surgery varies between 1% and 8%[28,29] and is mainly related to plaque dislodgment and peri-operative arrhythmia. Off-pump coronary surgery still requires manipulations of the ascending aorta for proximal vein graft anastomoses. Furthermore, it has not been quite successful in significantly decreasing post-operative atrial fibrillation.[30] This may explain why a clear consensus on the benefit of OPCAB surgery with regard to peri-operative stroke has not been reached yet.[15,31,32]

In our experience, the prevalence of stroke in OPCAB surgery has been comparable to conventional surgery.[15] This has been also the experience of the Northern New England Cardiovascular Disease Study Group.[32] Nevertheless, Cleveland et al[31] using the Society of Thoracic Surgeons National Adult Cardiac Surgery Database, compared the procedural outcome of 11,717 cases of OPCAB and 106,423 conventional surgeries. They found an absolute reduction of 2.1% (4.6% to 2.5%) in favour of OPCAB surgery, which reached statistical significance ($p<0.05$). Risk-adjusted mortality was comparable in both groups but OPCAB surgery resulted in an absolute 0.61% reduction of mortality, which was significant ($p<0.0001$). Avoiding aortic manipulation will certainly contribute to lowering the prevalence of peri-operative stroke. This aspect is discussed in Chapter 12.

Besides stroke, significant neurocognitive disturbances have been reported in up to 50% of the CABG population 6 weeks post-surgery.[33] Lloyd et al[34] in a randomized trial involving 60 patients undergoing off-pump versus on-pump surgery recorded a comparable neuropsychological outcome 12 weeks after surgery. Like others, they did not see a strong correlation between S-100β protein, a marker of Schwan's cell damage, and neuropsychological outcome.[35] Unfortunately, Loyd et al[34] only studied neuropsychological outcome preoperatively and at 12 weeks. Early post-operative assessment could have detected early differences between the 2 techniques. Diegler et al[36] from Leipzig, in a similar prospective, randomized study, found better neurocognitive performance 1 week after surgery in OPCAB patients. This improvement correlated with the decreased amount of high intensity transient signals recorded by transcranial Doppler during off-pump procedures. Lately, Van Djick et al[37] in a prospective and randomized study undertaken on 281 patients, reported results in favour of OPCAB 3 months ($p=0.03$) after surgery but with no significant improvement at 12 months ($p=0.09$). They also noted that stroke rate, mortality, and quality of life assessment were similar in both groups at 3 and 12 months. All patients included in the study were operated for the first time and were relatively young (61 years). Zamvar and colleagues,[38] who carried out a prospective, randomized trial on neurocognitive changes at 1 and 10 weeks, obtained similar findings favouring off-pump surgery.[38] It is likely, as time goes by, that the significance between both surgical techniques will disappear, since neurocognitive dysfunction tends to fade with time. Interestingly, some early neurocognitive impairment has been reported in up to 31% of patients undergoing peripheral vascular surgery prior to discharge. This suggests that CPB by itself is not accountable for all the post-operative neurological disturbances encountered after coronary surgery.[39] However, early post-operative impairment has been recognized as a predictor of long-term decline in neurocognitive performance, as demonstrated by Newman and colleagues in their eloquent work on longitudinal assessment of neurocognitive performance after coronary artery surgery.[40] They discovered that patients displaying an overall decline in the composite cognitive index after conventional coronary surgery manifested deterioration 5 years later, even though their performances had improved significantly at 6 months. Those who were not affected immediately after surgery maintained their performance at 6 months and 5 years. The latter obsrvation, if others confirm it in the future, may constitute a very strong plea in favour of OPCAB surgery.

Renal Physiology

Since the universal acceptance of CPB in cardiac surgery, it has been implicated as a major cause of post-operative renal insufficiency. Loss of pulsatile flow, hemodilution, hypothermia, and inflammatory reaction are all postulated to be contributing factors. It seems likely that avoiding CPB may prevent the deterioration of the renal function. Ascione et al[41] confirmed this in a very well-executed, prospective, randomized trial. Studying 50 patients, they demonstrated better-preserved glomerular filtration along with lower N-acetyl glucosidase excretion in the first 24 hours after OPCAB surgery. The latter is a biochemical marker of renal tubular function impairment. This beneficial effect has also been confirmed by others,[42] although some have expressed more criticism of it.[43] Most of these optimistic reports focused mainly on the early post-operative period and not on peak in-hospital deterioration. Maribel et al[43] in a nonrandomized, prospective study comparing 55 OPCAB patients to 635 CPB patients, could not establish that OPCAB grafting was an independent predictor of post-operative renal dysfunction. Increases in serum creatinine and its clearance were comparable regardless of the procedure used. However, their OPCAB cohort was quite small and selected. Furthermore, 25% of the OPCAB patients underwent a peri-operative angiography with contrast dye, compared to less than 1% among CPB patients. The latter could have contributed to the impairment of renal function in the OPCAB cohort. In our more recent experience, the acute degradation of renal function (creatinine increase of > 50 mmol/L) was less frequent in OPCAB compared to CPB.[44] In a previous survey focusing on patients with abnormal preoperative renal function (plasma creatinine level > 130 mmol/L), we observed comparable further renal deterioration with both surgical techniques.[45] Although the use of OPCAB surgery may help hamper the effects of ECC on renal function, other strategies will be necessary to optimize renal preservation in patients with preoperatively compromise renal function. According to recent encouraging experimental data, N-acetyl-L-cysteine pretreatment might be promising for the preservation of renal function during coronary surgery.[46]

Hypercoagulability

Hypercoagulability has been a poorly-explored topic in OPCAB surgery. The ECC is known to induce a fibrinolytic state.[47] However, to some extent, all types of surgeries, contribute to the expression of a pro-coagulable state.[48] Any surgical exposure involving bone fracture or partition is even more likely to predispose the patient to a pro-coagulable state due to the release of tissue factor, tissue plasmin activator and the local formation of thrombin-antithrombin complexes.[49] Midline sternotomy, a conventional approach in cardiac surgery, disturbs bone integrity and may predispose the patient to venous thrombo-embolic complications. Contrary to OPCAB surgery, full heparinization and the occurrence of a platelet dysfunctional state resulting from the extracorporeal circulation, as used in conventional surgery, may attenuate the prevalence of such complications. Early in our experience, we had a 1% rate of venous thrombo-embolic complications in OPCAB surgery.[50] This was not significantly higher than the rate observed in conventional surgery (0.5%) in our institution. However, OPCAB patients affected by these complications had less predisposing factors, which was worrisome. Since this observation, we adopted a universal prophylactic regimen of subcutaneous heparin, which has been successful. Further research in this field is warranted to establish the best prophylaxis in these patients.

Results of Randomized Studies

Three major prospective randomized studies have been published or presented comparing OPCAB and conventional coronary surgery with ECC. In 2001, the Netherlands Octopus group[22] published the results of a prospective, randomized trials carried on 281 patients.[22]

Both cohorts were comparable in terms of demographics and preoperative risk factors. Completeness of the coronary artery revascularization was comparable in both groups (83%). There was a 7.7% rate of conversion among the OPCAB patients. A relatively small number of grafts were performed (off-pump: 2.4±1.0, on-pump: 2.6±1.1) on patients who were mostly young (61.7+9.2 and 60.8+8.8 years). Major differences were a decreased incidence of blood product transfusion (3 vs. 13%), a lower CK-MB increment (by 40%, p<0.01), and a shorter hospital stay (by 1 day, p< 0.01) in the OPCAB cohort. The postoperative course up to 1 month and quality of life were comparable for both groups. The conclusion reached was that in a selected group of patients, OPCAB surgery provided similar cardiac outcome with evidence of better myocardial protection and the reduced use of blood products.

Angelini et al[23] reported the results of 2 combined, prospective, randomized trials (beating heart against cardioplegic arrest study, BHACAS 1 and 2). In BHACAS 1, the patients were highly selected. Those with recent myocardial infarction or circumflex disease were excluded. They were included in BHACAS 2, although patients with poor ejection fractions (<30%), supraventricular arrhythmia, previous CABG surgery, previous stroke, renal or respiratory impairment, and coagulopathy were excluded from both studies. Short-term and mid-term results (survival free from any cardiac-related events at 24 months) were comparable with both techniques. However, the use of inotropes, the incidence of blood transfusion, post-operative atrial fibrillation, chest infection, mechanical ventilation (>10 hours), ICU stay (>24 hours), and hospital stay (> 7 days) were lower in OPCAB patients.

In BHACAS 2, on-pump patients received tranexanic acid (2 g) before chest opening. This did not hinder the striking difference in transfusion rate observed between the 2 groups (off-pump: 13% vs. on-pump: 45%). The conclusions of these 2 studies were that off-pump coronary surgery reduced in-hospital morbidity without compromising mid-term outcome compared to conventional surgery with extra-corporeal assistance.

Puskas et al[51] reported their personal experience with 200 prospectively-randomized patients undergoing first-time coronary artery revascularization using both techniques. All patients had similar completeness of revascularization and comparable in-hospital and 30-day outcomes. As shown in the 2 previous studies, OPCAB patients had a shorter hospital stay along with reduced transfusion requirements and evidence of less myocardial injury. In contrast with previous reports, the average number of bypass grafts/patients was high (3.4 in each group), and the ratio of grafts performed/grafts intended was comparable (OPCAB: 1.00±0.18, CABG: 1.01±0.09).

All these studies are indubitably fine pieces of evidence suggesting that avoiding ECC could significantly improve the outcome of coronary revascularization. Although OPCAB surgery did not contribute to decreased operative mortality, it did significantly reduce the associated morbidity. The former is not a surprise if one takes into account the fact that those studies were carried out on "good risk", elective, first-time patients. The operative mortality is already extremely low in these patients, and huge numbers would be necessary to challenge this issue. Future studies, including "higher risk" patients, will be useful to explore the full benefit of the procedure to lower the mortality in this patient category.

Conclusion

Beating heart coronary artery surgery remains a new player in the large playground of coronary revascularization. It still has to acquire its "lettres de noblesse" to be fully accepted in the game. Nevertheless, new rules are progressively being written, and strategies redesigned. Definitely, off-pump surgery is part of the new surgical techniques that cannot be ignored any more by cardiac surgeons. It is likely that better comprehension of the pathophysiology of the ECC along with the development of new drugs, such as leukocyte depletion, anti-C5a complement, heparin-bound circuitry etc., will eventually decrease the-ECC generated inflammatory reaction in a significant fashion. This, however, will not be possible without extensive cost and research. In the meantime, current data have confirmed the efficiency of OPCAB surgery to overcome some of the side-effects of conventional surgery without jeopardizing the success of the procedure.

References

1. Kirklin JK, Westaby S, Blakestone EH et al. Complement and damaging effects of cardiopulmonary bypass. J Thorac Cardiovasc Surg 1983; 86:845-57.
2. Moat NE, Shore DF, Evans TW. Organ dysfunction and cardiopulmonary bypass: The role of complement and complement regulatory proteins. Eur J Cardiothorac Surg 1993; 7:563-73.
3. Duke T, South M, Stewart A. Altered activation of the L-arginine nitric oxide pathway during and after cardiopulmonary bypass. Perfusion 1997; 12:405-10.
4. Frering B, Philip I, Dehoux M et al. Circulating cytokines in patient undergoing normothermic cardiopulmonary bypass. J Thorac Cardiovasc Surg 1994; 108:636-41.
5. Edmunds Jr LH. Inflammatory response to cardiopulmonary bypass. Ann Thorac Surg 1998; 66:S12-6.
6. Kappelmayer J, Bertiabei A, Edmunds Jr LH et al. Tissue factor is expressed on monocytes during simulated extracorporeal circulation. Circ Res 1993; 72:1075-81.
7. Steinberg JB, Kapelanski DP, Olson JD et al. Cytokine and complement levels in patients undergoing cardiopulmonary bypass. J Thorac Cardiovasc Surg 1993; 106:1008-16.
8. Downing SW, Edmunds Jr LH. Release of vasoactive substances during cardiopulmonary bypass. Ann Thorac Surg 1992; 54:1236-43.
9. Brasil LA, Gomes WJ, Salomão R et al. Inflammatory response after myocardial revascularization with or without cardiopulmonary bypass. Ann Thorac Surg 1998; 66:56-9.
10. Gu YJ, Mariani MA, Oevewren WV et al. Reduction of the inflammatory response in patient undergoing minimally invasive coronary artery bypass grafting. Ann Thorac Surg 1998; 65:420-4.
11. Matata BM, Sosnowski AW, Galinānes M. Off-pump bypass graft operation significantly reduces oxidative stress and inflammation. Ann Thorac Surg 2000; 69:785-91.
12. Ascione A, Lloyd CT, Underwood MJ et al. Inflammatory response after coronary revascularization with or without cardiopulmonary bypass. Ann Thorac Surg 2000; 69:1198-204.
13. Pfister AJ, Zaki MS, Garcia JM et al. Coronary bypass without cardiopulmonary bypass. Ann Thorac Surg 1992; 54:1085-92.
14. Bouchard D, Cartier R. Off-pump revascularization of multivessel coronary artery disease has a decreased myocardial infarction rate. Eur J Cardiotorac Surg 1998; 14(Suppl 1):S20-4.
15. Cartier R, Brann S, Dagenais F et al. Systematic off-pump coronary artery revascularization in multivessel disease: Experience of three hundred cases. J Thorac Cardiovasc Surg 2000; 119:221-9.
16. Wan S, Izaat MB, Lee TW et al. Avoiding cardiopulmonary bypass in multivessel CABG reduces cytokine response and myocardial injury. Ann Thorac Surg 1999; 68(1):52-7.
17. Czerny M, Baumer H, Kilo J et al. Inflammatory response and myocardial injury following coronary artery bypass grafting with or without cardiopulmonary bypass. Eur J Cardiothorac Surg 2000; 17:737-42.
18. Rollins BJ. Chemokines. Blood 1997; 90:909-28.
19. Kukielka GL, Smith CW, La Rosa GJ et al. Interleukin-8 gene induction in the myocardium after ischemia and reperfusion in vivo. J Clin Invest 1995; 95:89-103.
20. Ivey CL, Williams FM, Collins PD et al. Neutrophil chemoattractants generated in two phases during reperfusion of ischemic myocardium in the rabbit: Evidence for a role for C5a and interleukin-8. J Clin Invest 1995; 95:2720-8.
21. Boyle Jr EM, Kovacich JC, Hebert CA et al. Inhibition of interleukin-8 blocks myocardial ischemia-reperfusion injury. J Thorac Cardiovasc Surg 1998; 116:114-21.
22. Van Dijk D, Niereich AP, Jansen EWL et al. Early outcome after off-pump versus on-pump coronary bypass surgery. Results from a randomized study. Circulation 2001; 104:1761-6.
23. Angelini GD, Taylor FC, Reeves BC et al. Early and midterm outcome after off-pump and on-pump surgery in Beating Heart Against Cardioplegic Arrest Studies (BHACAS 1 and 2): A pooled analysis of two randomized controlled trials. Lancet 2002; 359:1194-9.

24. Cox CM, Ascione R, Cohen AM et al. Effects of cardiopulmonary bypass on pulmonary gas exchange: A prospective randomized study. Ann Thorac Surg 2000; 69:140-5.

25. Taggart DP. Respiratory dysfunction after cardiac surgery: Effects of avoiding cardiopulmonary bypass and the use of bilateral internal mammary arteries. Eur J Cardiothorac Surg 2000; 18:31-7.

26. Lichtenberg A, Hagl C, Harringer W et al. Effects of minimal invasive coronary artery bypass on pulmonary function and postoperative pain. Ann Thorac Surg 2000; 70:461-5.

27. Ohkado A, Nakano K, Gomi A et al. The superiority of pulmonary function after minimally invasive direct coronary artery bypass. JPN J Thorac Cardiovasc Surg 2002; 50:66-9.

28. Roach GW, Kanchuger M, Mora M et al. Adverse cerebral outcomes after coronary bypass surgery. N Engl J Med 1996; 335:1857-63.

29. Mickleborough LL, Walker PM, Tagagi Y et al. Risk factors for stroke in patients undergoing coronary artery grafting. J Thorac Cardiovasc Surg 1996; 112:1258-9.

30. Siebert J, Rogowaki J, Jagielak D et al. Atrial fibrillation after coronary artery bypass grafting without cardiopulmonary bypass. Eur J Cardiothorac Surg 2000; 17:520-3.

31. Cleveland JC, Shroyer ALW, Chen AYC et al. Off-pump coronary artery surgery bypass grafting decreases risk-adjusted mortality and morbidity. Ann Thorac Surg 2001; 72(4):1282-9.

32. Hernandez F, Cohn WE, Baribeau YR et al. Northern New England Cadiovascular Disease Study Group. Ann Thorac Surg 2001; 72:1528-33.

33. Seines OA, Goldsborough MA, Borowicz LM et al. Neurobehavioral sequelae of cardiopulmonary bypass. Lancet 1999; 353:1601-6.

34. Lloyd CT, Ascione R, Underwood MJ et al. Serum S-protein release and neuropsychologic outcome during coronary revascularization on the beating heart: A prospective randomized study. J Thorac Cardiovasc Surg 2000; 119:148-154.

35. Westaby S, Saatvedt K, Katsumata T et al. Is there a relationship between serum S-100beta protein and neuropsychologic dysfunction after cardiopulmonary bypass. J Thorac Cardiovasc Surg 2000; 119:132-7.

36. Diegler A, Hirsh R, Schneider F et al. Neuromonitoring and neurocognitive outcome in off-pump versus conventional coronary bypass operation. Ann Thorac Surg 2000; 69:1162-6.

37. Van Djick D, Jansen EW, Hijman R et al. The Octopus Study Group. Cognitive outcome after off-pump and on-pump coronary artery bypass graft surgery: A randomized trial. JAMA 2002; 287:1405-12.

38. Zamvar V, Williams D, Hall J et al. Assessment of neurocognitive impairment after off-pump and on-pump techniques for coronary artery bypass graft surgery: Prospective randomized controlled trial. BMJ 2002; 325:1268-73.

39. Bates SPJ, Cartlidge NE, French JM et al. Neurologic and neuropsychological morbidity following major surgery: Comparison of coronary artery bypass and peripheral vascular surgery. Stroke 1987; 18:700-7.

40. Newman MF, Kirchner JL, Philips-Bute B et al. Longitudinal assessment of neurocognitive function after coronary artery bypass surgery. N Engl J Med 2001; 344:395-401.

41. Ascione R, Lloyd CT, Underwood MJ et al. On pump versus off pump coronary revascularization: Evaluation of renal function. Ann Thorac Surg 1999; 68:493-8.

42. Loef B, Henning R, Navis G et al. Beating heart coronary artery surgery avoids renal damage as compared with cardiopulmonary bypass [abstract]. Anesthesiology 1998; 89:A297.

43. Maribel G, Phillips-Bute B, Landolfo KP et al. Off-pump versus on-pump coronary artery bypass surgery and postoperative renal dysfunction. Anesth Analg 2000; 91:1080-84.

44. Cartier R. Current trend and technique in OPCAB surgery. J Cardiac Surg 2003; 18(1):32-46.

45. Cartier R. Off-pump surgery and chronic renal insufficiency. Letter to the Editor. Ann Thorac Surg 2000; 69:1995-6.

46. Conesa EL, Valero F, Nadal JC et al. N-acetyl-L-cysteine improves renal medullary hypoperfusion in acute renal failure. Am J Physiol Regul Integr Comp Physiol 2001; 281(3): R730-7.

47. Edmunds LH. Inflammatory response to cardiopulmonary bypass. Ann Thorac Surg 1998; 66:S12-6.

48. Dahl OE. Mechanism of hypercoagulability. Thromb Haemost 1999; 82:902-6.

49. Dahl OE. The role of the pulmonary circulation in the regulation of coagulation and fibrinolysis in relation to major surgery. Cardiothorac Vasc Anesth 1997; 11:322-8.

50. Cartier R, Robitaille D. Thrombotic complications in OPCAB surgery. J Thorac Cardiovasc 2001; 121(5):920-2.
51. Puskas JD, Williams WH, Duke PG et al. Off-pump coronary artery bypass grafting provides complete revascularization with reduced myocardial injury, transfusion requirements, and length of stay: A prospective randomized comparison of two hundred unselected patients undergoing off-pump versus conventional coronary artery bypass grafting. J Thorac Cardiovasc Surg 2003; 125(4):797-808.

CHAPTER 4

Principles of Stabilization and Hemodynamics in OPCAB Surgery

Raymond Cartier

S tabilization of the myocardium during coronary grafting remains a major task in OPCAB surgery. In the early beginning, only the anterior territory of the heart was targeted and acceptable rudimentary stabilization was achieved with myocardial stay sutures, pharmacological manipulations, and an adequately *synchronized* surgical assistant. However, these were not enough to provide safe access to the posterior wall of the heart, especially during circumflex artery stabilization. Considering that close to 75% of the current surgical patient population are referred for triple vessel revascularization, it became more than obvious that better means of stabilization had to be developed. During the last decade, several techniques and devices have been designed and used for myocardial stabilization during beating heart surgery. They are described in this chapter, along with a literature review on the topic.

What Is the Surgeon Looking for?

What would be, in the best circumstances, the ideal stabilizing set-up for surgical purposes? Clearly, surgeons want a bloodless and motionless field along with maximal degree of liberty to conduct anastomosis the same way they have been trained for under CPB conditions. So, stabilizers have to be "low profile", able to preserve heart motion as much as possible, and procure enough space for usual surgical instrument manipulation. Furthermore, they have to be used in association with a myocardium mobilization technique that does not create ventricular distortion and hemodynamic perturbations.

Current Techniques

Devices

Mechanical compression and suction devices are the most currently used stabilizing systems.[1-3] Both types of devices can procure satisfying stabilization to surgeons accustomed to them, although the best device for optimal posterior exposure is still being debated.[4] Compression devices have generally lower profile, which gives the surgeon better freedom of motion. Some of them allow the surgeon to secure the silicon loops directly on the stabilizer to avoid over-stretching the coronary arteries. On the other hand, suction devices stabilize with minimal manipulation, and have been shown not to influence coronary artery flow.[5] Occasionally, they can disconnect from the myocardial surface, especially in elderly patients on whom the epicardium is fatty and more friable. Life-threatening complications, such as intra-myocardial dissecting hematoma with epicardial rupture, have been described following suction stabilization.[6]

Off Pump Coronary Artery Bypass Surgery, edited by Raymond Cartier. ©2005 Eurekah.com.

But perhaps, what may be more important than the type of stabilizer used is the associated myocardial mobilization technique.

Myocardial Mobilization and Hemodynamics

Left Anterior Descending (LAD) Artery

The LAD artery is generally accessed with minimal manipulation. Either, few sponges or pericardial sutures anchored just above the left phrenic nerve are necessary to bring the apex and the anterior territory to the centre of the mediastinum. This is generally done with moderate hemodynamic disturbance. Manipulation of the anterior territory along with mechanical stabilization and coronary blood flow interruption causes a 6 to 15% decrease in cardiac output, which is generally well-tolerated.[7,8] Hemodynamics can also be affected during diagonal artery manipulations and, occasionally, with even a greater evidence of induced myocardial diastolic disease than LAD artery manipulation.[7] Nierich et al[8] in a clinical study, recorded a maximal drop in left ventricular stroke volume during diagonal artery manipulation in OPCAB surgery.[8] This was corrected with head-down positioning and the use of dopamine, which has also been our experience. In a study conducted on 50 patients with triple vessel disease undergoing complete revascularization, cardiac output dropped by 15% during both LAD artery and diagonal manipulation.[7] Moreover, the highest increase in pulmonary pressure occurred during diagonal artery grafting. Anatomically, the diagonal artery pathway crosses the left ventricular outflow tract, contrary to the LAD artery that is located over the septum. Any mechanical obstruction of the left ventricle outflow tract will increase left ventricular workload and decrease compliance of the ventricle. Manipulations of the diagonal territory have to be undertaken with delicacy and carefulness. Minimal load should be applied to achieve stabilization.

Severe hemodynamic instability can occur occasionally during manipulation of the anterior territory. Acute ischemia resulting from coronary snaring, in cases of long LAD artery contouring the apex or a poorly collateralized LAD artery with moderate stenosis, can temporarily cause severe anterior wall akinesia. Intra-coronary shunting is mandatory in these circumstances to alleviate the blood flow obstruction and restore ventricular contractility.

Posterior Wall Stabilization, Pericardial Traction

Early in our experience, we adopted the use of compression stabilizers and deep pericardial stay sutures for posterior wall exposure, methods that were developed independently by several individuals.[3,8-10] Four pericardial tractions (Ticron 00) are generally anchored between the left superior pulmonary vein and the inferior vena cava (IVC), very close (less than 2 cm) to the pericardial reflection at the base of the heart. Verticalization of the apex is then achieved in a majority of cases without any further manipulations. Special care should be taken during this maneuver to avoid injuring the lung parenchyma. The anesthesiologist should always temporarily deflate the lungs during passage of the needle. This helps decrease the occurrence of pneumothorax after surgery. The stabilizer is then applied to the targeted marginal branches. In previous reports, we have documented that adequate stabilization and grafting can be achieved with an average cardiac index fall of 10% and a mean pressure decline of 15%.[11] Head-down tilt of 20 to 30% was used in all patients, and a vasopressor, mainly an α-agonist, was given in close to 60% of them to maintain sufficient hydrostatic pressure and improve right ventricular inflow. Experimentally in large animals, Gründeman et al[12] have shown that verticalization alone without pericardial traction and head-down tilt caused a 32% drop in cardiac output and a 44% decrease in stroke volume. In their experiment, the heart was verticalized by pulling directly on the ventricle wall by suction. The direct tension applied to the posterior wall of the ventricle could have interfered with ventricular geometry and contractility, contributing to the hemodynamic perturbations observed. With the Bristol technique, Watters et al[13] snared a half-fold swab to the posterior pericardium halfway between the IVC and the left inferior pulmonary vein to mobilize the posterior wall. The extremities of the swab were subsequently applied to the anterior surface of the heart to help stabilize the posterior wall. Considerable wall restriction could be transferred to the right as well as the left heart during this maneuver because of the circumferential constraint (the swab in front and the right pericardium in the back) generated around the myocardium with this technique. Subsequent application of the stabilizer can only contribute to further increasing wall motion. In the 29 consecutive patients they studied with this technique, a 25% decrease in cardiac output was observed in addition to a significant rise in pulmonary artery pressure. In their experience, the circumflex territory was the most affected during revascularization. Matison et al[14] from Dallas, Texas, also reported significant changes during positioning.[14] They monitored both right and left atrial and ventricular end-diastolic pressures in 44 patients. Cardiac flow was recorded with an ultrasound transit-time flow probe. They also found maximum hemodynamic perturbation during posterior wall mobilization. Compared to the anterior territory, mean arterial pressure and stroke volume dropped by 22% and 29% respectively during posterior positioning instead of 4% and 18%. Although they did not specify their mobilization technique for the posterior wall, they mentioned that the heart was suspended in a pericardial cradle and a suction tissue stabilizer was applied. The head-down tilt position was systematically utilized during posterior positioning. An interesting aspect of their study was the variation in right heart pressure during positioning. A very significant elevation of right ventricular end-diastolic pressure occurred in all territories. They recorded increments of 151% and 67% during posterior wall and anterior wall positioning respectively. A similar but lower rise in atrial pressure was also noted. They concluded that severe right ventricular diastolic disease was induced during myocardial mobilization and coronary stabilization. This could be largely responsible for the hemodynamic instability during positioning, since it interferes with right ventricular preload. It could be partialy compensated by head-down tilt, which increases hydrostatic pressure, and by the administration of an α-agonist that pharmacologically produces the same effect.

Posterior Wall Stabilization, Apical Suction

Apical suction devices have been developed to assist posterior exposure and replace pericardial traction sutures. The Starfish (Metronic, Minneapolis, MN) and the X*pose* devices (Guidant Corporation, Cupertino, CA) are the most popular ones.[15,16] They allow verticalization of the heart by directly mobilizing the apex, instead of the base of the ventricle as in the pericardial technique. This technique eases the procedure of mobilization since very little manipulation of the heart is necessary. On the other hand, there is potential for ventricular distortion since a direct force is applied to the wall before the stabilizer. Occasionally, disconnection can occur during anastomotic time with potential injury to the grafted vessel. Very little literature is available on hemodynamic evaluation with this technique of mobilization, but it has gained considerable popularity lately. Direct traction on the apex could contribute to heart remodeling and ultimately decreasing right ventricular diastolic disease.[17] Sepic et al[18] in a very elegant experimental study in the porcine model, have shown that hemodynamics impairment can be avoided by apical suction. This, however, was recently challenged in a in a clinical trial. Gummert et al[19] propectively evaluated this issue in 27 patients undergoing back-wall revascularization. Exposure was obtained either by pericardial stitches or apical suction in every patient. On average, 2.9±0.7 anastomosis per patient were performed. Changes in mean arterial pressure and cardiac index were comparable for both lateral and inferior wall as well as were variations for other hemodynamic parameters. The author concluded that, for back-wall exposure, apical suction did not offer any additional benefit over the pericardial stitches.

Posterior Descending and Right Coronary Artery

The inferior wall is generally accessed with minimal surgical mobilization. We found it particularly helpful to anchor 1 or 2 pericardial tractions just above the IVC to help rotate the base of the heart towards the left. This brings the distal right coronary artery directly on the midline. By adding 20-degree head-down tilt, the surgeon gets good exposure to the distal right coronary artery. The posterior descending artery is accessed by setting the patient in Trendelenburg positioning, after mobilization of the apex in the midline either with pericardial stitches or apex suction. Hemodynamics are generally only moderately affected during these maneuvers.

Systolic and Diastolic Function during OPCAB Surgery

Diastolic function is a marker of ischemia and a common feature in coronary insufficiency. By relieving ischemia, revascularization is expected to improve diastolic dysfunction. Paradoxically, diastolic dysfunction has been reported after conventional coronary revascularization.[20] The duration of ischemic arrest, reperfusion injury, hemodilution, free radical generation during cardiopulmonary bypass (CPB), leukocyte activation, lipid peroxidation, hypothermia, rapid pH changes, and myocardial edema are all recognized as contributing factors to diastolic dysfunction during conventional coronary surgery.[21-24] Although the introduction of cardioplegic arrest has contributed to improve systolic and diastolic function,[25,26] abnormal diastolic function has still been reported after conventional cardiac surgery,[27] regardless of the type of cardioplegic solution used. Diastolic dysfunction can occur even in patients with normal coronaries and seems to be associated with subendocardial ischemia.[28] McKenney et al[29] assessed the incidence and severity of left ventricular diastolic dysfunction immediately after conventional coronary artery bypass surgery. They undertook hemodynamic and transesophageal echocardiographic evaluation of 20 patients undergoing bypass surgery. To study the diastolic dysfunction, they specifically looked at the relationship between pulmonary capillary wedge pressure and the left ventricular (LV) end diastolic area. Compared to baseline, all patients displayed a decreased LV end diastolic area after bypass surgery at each pressure level, confirming the increased stiffness and reduced compliance of the LV (Fig. 1). The deceleration time of mitral inflow, along with early peak velocity/late peak auricular velocity, which are echocardiographic markers of diastolic dysfunction, were significantly decreased after surgery. The authors concluded that diastolic dysfunction, as evaluated by end diastolic LV stiffness and mitral valve deceleration time, remained a very common feature after coronary bypass surgery.

Global ischemia and reperfusion injury are avoided during OPCAB surgery. However, stabilizers rigidify the ventricular wall, and the concomitant blood flow interruption could cause temporary akinesia of the territory involved. Biswas et al[30] from Duke University have shown that systolic and diastolic functions are affected during OPCAB surgery. They assessed the systolic wall motion score index for systolic function and mitral inflow velocity for diastolic function. Maximal systolic function worsening occurred during circumflex grafting although some also occurred during LAD artery grafting (Fig. 2). Diastolic dysfunction peaked during circumflex wall grafting (Fig. 3). The surgical technique they used combined 2 pericardial traction sutures and a suction type of stabilizer to access the posterior wall. Contrary to previous reports on CPB, systolic and diastolic dysfunction disappeared with removal of the stabilizer and returned to the normal heart position. No residual dysfunction was observed. As we noted in our study on OPCAB hemodynamics, the cardiac output variations recorded during grafting were similar regardless of the territory grafted, and were not circumflex-specific. The authors concluded that even though significant diastolic and systolic dysfunction occurred during circumflex grafting, it was a temporary feature and should not preclude surgeons from considering patients with severe circumflex artery disease.

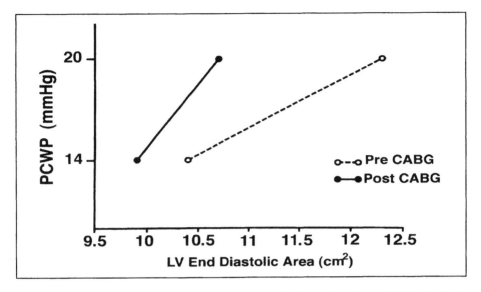

Figure 1. Pressurearea relationship in a representative patient before (Pre) and immediately after (Post) coronary artery bypass surgery (CABG). Baseline pulmonary capillary wedge pressure (PCWP) of 14 mm Hg is increased with volume loading. After bypass surgery, the left ventricular (LV) end diastolic area is smaller at each pressure level, as reflected by a leftward shift of the pressurearea relationship. Reprinted from McKenney et al. J Am Coll Cardiol 1994; 24:1189-94 with permission.

Adjunctive Techniques

Trendelenburg Positioning

Many adjunctive techniques have been described to optimize and facilitate myocardial exposure. As mentioned earlier in this chapter, head-tilt positioning, also called Trendelenburg positioning, is the most popular. Frederich Trendelenburg introduced the head-down position in 1890 to facilitate the surgical exposure of pelvic organs.[31] Although its hemodynamic effects on carotid flow and blood pressure have been questioned in experimental shock situations[32,33] its impact on humans has been more predictable.[34,35] Deklunder et al[34] studied the effect of head-down tilt (HDT) in 12 healthy male subjects aged 19-24 years old. HDT promptly induced an 8% increase of trans-aortic flow velocity that translated to a 6% elevation of cardiac output with 7 and 15% augmentation of diastolic and systolic systemic pressures respectively. These hemodynamic features were maintained for the 5-minute duration of the experiment. The authors concluded that the cardiac response to HDT positioning was mainly due to passive changes in ventricular filling, secondary to blood shift rather than autonomous nervous regulation. The preload effect of HDT has been shown to be preserved in the elderly.[36] Some concerns have been expressed about the cerebral circulation and potential cerebral edema during HDT positioning.[37] Numerous studies have reported that autoregulation and cerebral metabolic requirements were generally preserved during Trendelenburg positioning, even up to an HDT of 75 degrees.[38,39] There is also evidence that even patients with autonomic nervous system failure have preserved cerebral autoregulation during position changes.[40] Trendelenburg positioning during OPCAB surgery contributes to increase venous hydrostatic pressure and passive filling of the right ventricle in the presence of acquired diastolic dysfunction during apex verticalization and posterior stabilization. Its utilization is safe, simple, and helpful in assisting the surgeon during exposure of the circumflex territory.

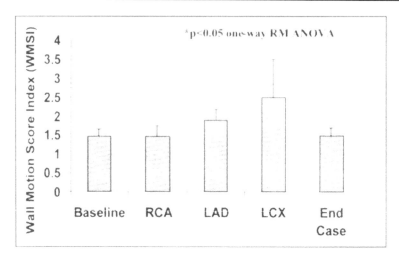

Figure 2. Wall motion score index (WMSI) at baseline and 10 min into grafting of the right, left anterior descending and left circumflex coronary arteries, and at the end of each operation. Reprinted from Biswas et al. Eur J Cardiothorac Surg 2001; 20:913-17 with permission.

Right Tilt

Besides HDT positioning, rightward rotation of the table (towards the surgeon) is also commonly undertaken during posterior exposure. Grundeman et al[43] demonstrated in the porcine model that hemodynamics were maintained during this maneuver. A 60-degree turn to the right significantly increased mean arterial pressure, right atrial pressure, and right ventricular end diastolic pressure, whereas left ventricular preload was maintained. Associated with verticalization, the maneuver greatly improves surgical exposure.

Herniation and Pericardial Division

"Herniation" of the heart is also a technique described to optimize visualization of the posterior wall during circumflex grafting.[42] It consists of dividing the left pericardium underneath the phrenic nerve. The purpose is to create space to allow the apex of the heart to shift freely in the left chest cavity. Surgical working space is improved, and less restriction is applied to the right ventricle.[28] Torsion of the myocardium as it falls in the pleural cavity can impede venous return by distorting the vena cava and causing unstable hemodynamics.[42] It should be applied carefully. A variant of this maneuver consists of vertical division of the right pericardium at the phreno-pericardial junction. This division can be further extended upward towards the inferior pulmonary vein, increasing space for the right ventricle and also deepening the base of the heart. The surgical working space on the posterior wall is consequently enhanced. We have found this maneuver to be particularly helpful in the enlarged heart where the apex protrudes more extensively than usual during verticalization.[43] By deepening the base of the heart, apex protrusion is decreased, and surgical working space on the back of the heart is increased. This will be further discussed in Chapter 6.

IVC Snaring

Occasionally, heart mobilization and regional ischemia during coronary blood flow interruption cause sudden pulmonary hypertension. This is generally due to ischemic left ventricular dysfunction and temporary ischemic mitral insufficiency.[44] Although they could occur in all territories, they appear to be more common during grafting of the anterior territory according

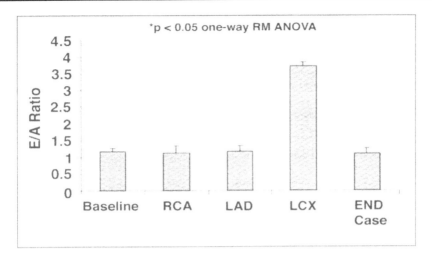

Figure 3. Early peak velocity/Late peak auricular velocity ratio at baseline and 10 min into grafting of the right, left anterior descending and left circumflex coronary arteries, and at the end of each operation. Reprinted from Biswas et al. Eur J Cardiothorac Surg 2001; 20:913-7 with permission.

to our experience. An easy way to reverse these features is by temporarily interrupting blood flow of the IVC. This can be attained either by cross-clamping or snaring of the IVC. The interruption can be partial or total, and can generally be maintained for a few minutes. We have currently used the technique in close to 10% of our cases. By temporarily decreasing venous preload, IVC-snaring lowers the filling pressure of the right and left ventricles and rapidly improves hemodynamics. Pulmonary artery pressure returns to baseline, as does left ventricle filling pressure. The left ventricular distention is reversed, which attenuates the mitral regurgitation that normally is triggered by ischemia.[45] The maneuver gives the surgeon time to complete the anastomosis, and provides the anesthesiologist with freedom to optimize the management of nitroglycerine. Ultimately, emergent conversion can be avoided.

Conclusion

Myocardial stabilization has been greatly improved since 1995. Refinement of coronary stabilizers and new ways to mobilize the heart have contribute considerably to this accomplishment. Better comprehension of physiology and hemodynamics during these manipulations, along with enabling technologies, should help ease the surgeon's ability to accomplish coronary revascularization in the near future.

References

1. Borst C, Jansen EWL, Grundeman PF et al. Regional cardiac wall immobilization for open chest and closed chest artery bypass grafting in the beating heart: The "Octopus" method [abstract]. Circulation 1995; 92:I-177.
2. Baumgartner FJ, Gheissari A, Capouya ER et al. Technical aspects of total revascularization in off-pump coronary bypass via sternotomy approach. Ann Thorac Surg 1999; 67:1653-8.
3. Cartier R, Brann S, Dagenais F et al. Systematic off-pump coronary artery revascularization in multivessel disease: Experience of three hundred cases. J Thorac Cardiovasc Surg 2000; 119:221-9.
4. Amano A, Osaki S, Nagano N et al. New subjects for exceeding conventional on-pump CABG. Kyobu Geka 2001; 54:262-9.
5. Grundeman PF, Coreelius B, van Herwaarden JA et al. Vertical displacement of the beating heart by the Octopus tissue stabilizer: Influence on coronary flow. Ann Thorac Surg 1998; 65:1348-52.

6. Mandke NV, Nalladaru ZM, Chougule A et al. Intra-myocardial dissecting hematoma with epicardial rupture, an unusual complication of the octopus 3 stabilizer. Eur J Cardiothoracic Surg 2002; 21:566-7.

7. Do Q-B, Goyer C, Chavanon O et al. Hemodynamic changes during off-pump CABG surgery. Eur J Cardiothorac Surg 2002; 21:385-90.

8. Nierich AP, Diephuis J, Jansen EWL et al. Heart displacement during off-pump CABG: How well is it tolerated. Ann Thorac Surg 2000; 70:466-72.

9. Baumgartner FJ, Gheissari A, Capouya ER et al. Technical aspects of total revascularization in off-pump coronary bypass via sternotomy approach. Ann Thorac Surg 1999; 67:1653-8

10. Bergsland J, Schimd S, Yanulevich J et al. Coronary artery bypass grafting (CABG) without cardiopulmonary bypass (CPB): A strategy for improving results in surgical revascularization. Heart Surgery Forum 1998; 1:107-110.

11. Do QB, Cartier R. Hemodynamic changes during beating-heart CABG surgery (in French). Ann Chir 1999; 53:706-11.

12. Gründeman PF, Borst C, van Herwaarden JA et al. Hemodynamic changes during displacement of the beating heart by the Utrecht Octopus method. Ann Thorac Surg 1997; 63:S88-92.

13. Watters MPR, Ascione R, Ryder IG et al. Haemodynamic changes during beating heart coronary surgery with the "Bristol Technique". Eur J Cardiothorac Surg 2001; 19:33-40.

14. Matison M, Edgeton JR, Horswell JL et al. Analysis of hemodynamic changes during beating heart surgical procedures. Ann Thorac Surg 2000; 70:1355-60.

15. Niinami H, Takeuchi Y, Ichikawa S et al. Multivessel off-pump coronary aortic bypass grafting with an impaired and severely dilated left ventricle using the Starfish heart positioner. Kyobu Geka 2002; 55(9):773-7.

16. Dullum MK, Resano FG. Xpose: A new device that provides reproducible and easy access for multivessel beating heart bypass grafting. Heart Surg Forum 2000; 3(2):113-7.

17. Mach M. Latest technique in OPCAB surgery. Presented at Peak Performance. Lake Louise, Alberta: February 28th 2002.

18. Sepic J, Wee JO, Soltesz EG et al. Cardiac positioning using an apical suction device maintains beating heart hemodynamics. Heart Surg Forum 2002; 5(3):279-84.

19. Gummert JF, Raumanns J, Bossert T et al. Suction device versus pericardial retraction sutures. Comparison of hemodynamics using different exposure systems in beating heart coronary surgery. Heart Surg Forum 2003; 6(Suppl 1):S32.

20. Lawson WE, Seifert F, Anagnostopoulos C et al. Effect of coronary artery bypass grafting on left ventricular diastolic function. Am J Cardiol 1988; 61:283-7.

21. Spotnitz HM, Bregman D, Bowman FO et al. Effects of open heart surgery on end-diastolic pressurediameter relations of the human left ventricle. Circulation 1979; 59:662-71.

22. Apstein CS, Lorell BH. The physiologic basis of left ventricular diastolic dysfunction. J Cardiovasc Surg 1988; 3:475-85

23. Prasad K, Kalra J, Bharadwaj B et al. Increased oxygen stress free radical activity in patients on cardiopulmonary bypass undergoing aortocoronary bypass surgery. Am Heart J 1992; 123:3-45.

24. Ivanov J, Wiesel RD, Mickleborough LL et al. Rewarming hypovolemia after aortocoronary bypass surgery. Crit Care Med 1984; 12:1049-54.

25. Ellis RJ, Mangano MD, Van Dyke DC et al. Hypothermic potassium cardioplegia preserves myocardial compliance. Surgery 1979; 86:810-7.

26. Natsuaki M, Itoh T, Ohteki H et al. Evaluation of left ventricular early diastolic function after coronary artery bypass grafting relating to myocardial damage. Jpn Circ J 1991; 55:117-24.

27. Weng Z, Nicolosi AC, Detwiler PW et al. Effects of crystalloid, blood, and University of Wisconsin perfusates on weight, water content, and left ventricular compliance in an edema-prone, isolated porcine heart model. J Thorac Cardiovasc Surg 1992; 103:504-13.

28. Buckberg GD, Toweres B, Paglia DE et al. Subendocardial ischemia after cardiopulmonary bypass. J Thorac Cardiovasc Surg 1972; 64:669-80

29. McKenney PA, Apstein CS, Mendes LA et al. Increased left ventricular diastolic chamber stiffness immediately after coronary artery bypass surgery. J Am Coll Cardiol 1994; 24:1189-94.

30. Biswas S, Clements F, Diodato L et al. Changes in systolic and diastolic function during multivessel off-pump coronary bypass grafting. Eur J Cardiothorac Surg 2001; 20:913-17.

31. Trendelenburg F. Ueber Blasenscheidenfisteloperationen und ueber Beckenhochlagerung bei operationen in der Bauchhoehle. Samml Klin Vortrage 1890; 355(109):3373-92.
32. Speert H. Friedrich trendelenburg and the trendelenburg position. Surg Gynecol Obstet 1957; 105(1):114-9.
33. Guntheroth WG, Abel FL, Mullins GL. The effect of Trendelenburg's position on blood pressure and carotid flow. Surg Gynecol Obstet 1964; 119:345-8.
34. Deklunder G, Lecroart JL, Chammas E et al. Intracardiac hemodynamics in man during short periods of head-down and head tilt. Aviat Space Environ Med 1993; 64(1):43-9.
35. Mengesha YA. Comparative study of haemodynamic response to active posture inducing head-ward pooling o blood in man. East Afr Med J 2001; 78(1):212-5.
36. Voutilainen S. Effects of head-up and head-down tilt on the transmitral flow velocities in relation to age: A Doppler echocardiographic study in healthy persons. Clin Physiol 1994; 14(5):561-7.
37. Hu Z, Zhao G, Xiaio Z et al. Different responses of cerebral vessels to −30 degrees head-down tilt in humans. Aviat Space Environ Med 1999; 70(7):674-80.
38. Katkov VE, Chestukhin VV, Rumiantsev VV et al. Pressure in the jugular vein and right atrium and the cerebral hemodynamics of the healthy human being exposed to postural effects. Kosm Biol Avakosm Med 1981; 15(5):49-53.
39. Terai C, Anada H, Matsushima S et al. Effects of mild Tredelenburg on the central hemodynamics and internal jugular velocity, cross-sectional area, and flow. Am J Emerg Med 1995; 13(3):255-8.
40. Bondar RL, Dunphy PT, Moradshahi P et al. Cerebrovascular and cardiovascular responses to graded tilt in patients with autonomic failure. Stroke 1997; 28(9):1677-85.
41. Grundeman PF, Borst C, Verlaan CW et al. Hemodynamic changes with right lateral decubitus body positioning in the tilted porcine heart. Ann Thorac Surg 2001; 72(6):1991-6.
42. Novitzky D, Boswell BB. Total myocardial revascularization without cardiopulmonary bypass utilizing computer-processed monitoring to assess cerebral perfusion. Heart Surg Forum 2000; 3(3):198-202.
43. Cartier R. "Tips and tricks in beating heart surgery". Invited lecturer. 2002 Adult Cardiac Symposium, Annual Congress of the American Association for Thoracic Surgery. Washington, DC: March 2002.
44. Dagenais F, Cartier R. Pulmonary hypertension during beating heart coronary surgery: Intermittent inferior vena cava snaring. Ann Thorac Surg 1999; 68:1094-95.
45. Couture P, Denault AY, Sheridan P et al. Partial inferior vena cava snaring to control ischemic left ventricular dysfunction: [Constriction partielle de la veine cave inferieure pour controler une dysfonction ventriculaire gauche]. Can J Anaesth Apr 2003; 50(4):404-10.

CHAPTER 5

Anesthetic Management for OPCAB

Patrick Limoges, Robert Blain and Peter Sheridan

Introduction

On October 16, 1846, William T.G. Morton initiated the modern era of anesthesia. The developments that followed the first use of ether would permit the realization of increasingly complex surgical procedures, such as cardiac revascularization.

Since the introduction of myocardial revascularization surgery in 1953, most procedures have been performed under cardiopulmonary bypass (CPB). However, in the early 1990s, several factors led to renewed interest in OPCAB. Technical and pharmacological advances in interventional cardiology made the treatment of angina more efficient. The reduced cost and relatively low mortality and morbidity of these techniques rendered the surgical option less appealing. CPB is known to trigger an inflammatory response. For most patients, the consequences of this inflammatory response are benign, but for a minority, the systemic effects are not negligible; being deleterious, they lead to increased mortality and morbidity.

All major surgeries are associated with an inflammatory response, and off-pump coronary revascularization is no exception. Nevertheless, studies have shown lower levels of inflammation markers in patients when CPB was avoided.[1,2] This diminution of the inflammatory response seems to translate into less organ dysfunction, a reduced need for post-operative inotropes, a tendency towards decreased morbidity, a diminution of transfusion, shorter hospital stays, and lower total costs.[3,4]

Surgical candidates are becoming older, and present more frequently with multiple risk factors. Although advances in the fields of anesthesia, surgery and perfusion have allowed higher-risk patients to undergo cardiac surgery under CPB with no increased mortality,[5] lower morbidity has been noted in OPCAB patients.[3,6]

In this chapter, we will specifically cover anesthetic management during revascularization on the beating heart via median sternotomy. Anesthetic considerations specific to minimally invasive direct coronary artery bypass (MIDCAB) or endoscopic and endovascular approaches (port access) will not be discussed.

Pre-Operative Period

The pre-operative evaluation of patients undergoing OPCAB does not differ significantly from the routine pre-operative assessment of patients submitting to major surgery. It is important to identify health problems associated with coronary artery disease that may have an impact on anesthetic management. Accordingly, systemic or pulmonary hypertension, smoking history, diabetes, bleeding disorders, and renal or hepatic insufficiency, must be taken into account. Anesthetic history, steroid dependence, allergies and current medications are also relevant. During pre-operative evaluation, the anesthesiologist should note not only the number

---

of vessels affected, but also the location and severity of the coronary artery stenoses. Armed with this knowledge, it is possible to anticipate the hemodynamic changes that may occur with surgery.

Currently, cardiac medications are prescribed for the morning of surgery with the exception of diuretics. Whether or not angiotensin-converting enzyme inhibitors or angiotensin II receptor antagonists should be withheld on the morning of surgery is controversial. A possible link between these drugs and persistent hypotension after the induction of anesthesia and while on CPB has led some authors to strongly recommend omitting them on the day of surgery.[7,8] Others have not found any related hypotensive effect but rather increased post-operative vasodilator requirements in patients when the morning dose was withheld.[8,9] Although no specific study has been conducted on this specific topic in OPCAB patients, it is our practice to refrain from giving these drugs on the day of surgery.

Premedication

Premedication aims to alleviate anxiety and its adverse effects such, as hypertension and angina. It also makes the patient more comfortable during the placement of intravascular catheters under local anesthesia. Combination of a narcotic (morphine 0.1-0.15 mg/kg IM) and a sedative (scopolamine 5 μg/kg IM or midazolam 0.05 mg/kg IM) is common practice. In unstable or debilitated patients, the premedication dose is reduced or omitted. Supplemental oxygen is administered after the patient is premedicated to reduce the possibility of arterial oxygen desaturation.

Intra-Operative Period

In general terms, the anesthetic for an OPCAB procedure is similar to that used for coronary artery bypass grafting (CABG) surgery on CPB. However, fluid replacement strategies, treatment of hemodynamic instability, and the prevention of hypothermia are particular to OPCAB operations and will be discussed below. Meticulous attention to these areas is crucial to the success of the procedure.

Monitoring the OPCAB Patient

In addition to routine intra-operative monitoring that applies for any anesthetic,[9] a 5-lead ECG is deployed to continuously display leads II and V5. This two-lead combination has the highest sensitivity for the detection of myocardial ischemia.[10] Monitoring for myocardial ischemia relies on ST segment analysis. Although manipulation of the heart can affect ST segments and QRS voltage, it is assumed that any changes from baseline after the heart has been mobilized for bypass surgery are due to ischemia. The ST segment trend over the preceding 30 minutes is displayed, facilitating the detection of ischemia. A large bore IV and an arterial pressure monitoring line are secured. A central line is placed either before or after the induction of anesthesia at the discretion of the anesthesiologist. Finally, at the Montreal Heart Institute, a pulmonary artery catheter is considered useful, particularly to guide fluid therapy in patients with left ventricular dysfunction[11] and to manage hemodynamic instability in the operating theatre and the intensive care unit (ICU).

Trans-esophageal echocardiography (TEE) is not performed routinely, similarly to our practice for CPB cases. The role of TEE will be discussed in another chapter, but some salient points will be mentioned here. For a CABG on CPB, TEE is a category II indication, meaning that there is no evidence that it changes the outcome.[12] In a retrospective analysis of 703 patients, Couture et al[13] showed that TEE changed surgical management in only 5% of patients undergoing coronary revascularization. It may be indicated for the evaluation of other lesions, such as mitral insufficiency or aortic atheromatous disease, that can alter the surgical plan. Although not as sensitive as epiaortic scanning in the detection of atheromatous lesions of the thoracic aorta, TEE can be valuable in patients at high risk for neurological complications.[14]

When TEE is performed, the baseline exam should be conducted before the surgical incision is made. Then, quality images free of ischemia related to the procedure can be recorded. Once surgical manipulations have begun, the transgastric view may be difficult or impossible to obtain as the apex of the heart is elevated for distal anastomoses and a sponge is put behind the heart. For this reason, the mid-esophageal view behind the left atrium is favoured because it generally allows imaging of all myocardial wall segments with the 4-chamber, 2-chamber and long-axis planes. Zones of apparent akinesia or dyskinesia related to the cardiac stabilizer system may be detected. The preservation of myocardial thickening in this scenario, allows mechanical compression of the heart to be distinguished from myocardial ischemia. Naturally, the effects of mechanical compression will vanish once the stabilizing system is released. New wall motion abnormalities (NWMA) occurring after revascularization are worrisome and may reflect inadequate revascularization. Desirable as it is, a return to baseline does not guarantee a perfectly permeable bypass. Although TEE is more likely to be used in patients with decreased systolic function, NWMA may be harder to detect in that group.

Prevention of Hypothermia

Care must be taken to prevent hypothermia, particularly if early extubation is planned. The anesthetized patient is essentially a poikilotherm, and active measures are required to ensure that heat loss does not exceed heat gain. It is obviously easier to maintain normothermia than to undertake active rewarming of the patient. The adverse effects of hypothermia on wound healing, coagulation and arrhythmia are well-established.[15,16] In a study of noncardiac surgical patients, Frank et al[17] observed an increased incidence of ischemic events in those exposed to post-operative hypothermia. Prevention of hypothermia requires a combination of several actions. Heat loss is proportional to the period of time that the anesthetized patient is left uncovered; this interval should be reduced to the minimum. Operating room temperature is normally set at 21°C. A heating mattress set at 40°C is placed under the patient. A fluid warmer is applied to all intravenous fluids (Hotline Model HL90, Level 1 Technologies, Rockland, MA). The saline solutions for irrigation of the operative field are prewarmed. Respiratory heat losses can be reduced by using, low, fresh gas flow and by connecting the endotracheal tube to a heat and moisture exchanger. In case of small body surface patients, a forced air-warming blanket (Warm Touch Model 5100, Mallinckrodt Medical, St. Louis, MO) is positioned over the patient's head and shoulders to optimize body heat preservation.

Induction of Anesthesia

As for conventional CABG, the anesthetic regimen aims to be compatible with the goal of early extubation. Drugs of short or intermediate duration of action are chosen. Several drug combinations can be given for this purpose.[18] There is no specific reason to chose one drug combination over another as long as the goals of hypnosis, analgesia, amnesia, autonomic stability and neuromuscular block are achieved in a manner that is compatible with early extubation. Several narcotics can help to tolerate the intubation and reduce pain associated with the surgical procedure. Fentanyl 15-50 μg/kg, sufentanyl 1.5-3.0 μg/kg and even morphine 0.5-1.5 mg/kg are commonly used in our institution. These doses are 4-10 times lower than those administered in the late 1970s.[19,20] Remifentanil, a potent ultra short-acting narcotic, can be delivered at an infusion rate of 0.15-0.4 μg/kg/min. Unconsciousness is generally induced with one of the following: pentothal 2-3 mg/kg, midazolam 0.1-0.2 mg/kg, or propofol 1-2 mg/kg. In addition to the above-mentioned narcotics, anesthesia is maintained with propofol infusion or a volatile agent, such as isoflurane or sevoflurane.

Inhalation agents can be of particular interest by virtue of their association with ischemic preconditioning. Although the mechanism underlying the phenomenon needs further elucidation, it could be related to adenosine, protein kinase C (PKC) and K_{ATP} channel metabolism.[21] Isoflurane activates K_{ATP} channels, and has been shown to induce preconditioning in animal and human models.[22-25] Belhomme et al[26] in a study performed on 20 patients undergoing CABG with CPB, treated 10 of them with 2.5 MAC isoflurane for 5 minutes before coronary anastomoses were implemented. Creatine phosphate kinase (CK-MB) fractions and troponin-I were significantly lower in the isoflurane-treated group suggesting better myocardial preservation with this drug. Among other commonly-used agents that can stimulate the pathways related to preconditioning are beta-agonists and exogenous calcium through the activation of PKC.[20]

Neuromuscular blocking is generally achieved with drugs of medium (rocuronium, vecuronium or cisatracurium) or long (pancuronium) duration of action. To permit fast tracking, residual neuromuscular blockade is normally antagonized at the end of the procedure.

During the period between anesthesia induction and the beginning of coronary artery anastomoses, 2-5 g of magnesium sulfate are administered. It has been suggested that pre-operative magnesium delivery diminishes the prevalence of post-operative atrial tachyarrhythmias.[27-29] Furthermore, hypomagnesemia is commonly observed in the peri-operative period in our cases. As an intracellular cation, serum magnesium does not accurately reflect total body stores. A low-magnesium plasmatic level is normally associated with a large deficiency. Magnesium can also decrease arterial graft spasm.[30] Supplements rarely cause hypotension when administered slowly and can prolong the duration of action of neuromuscular blocking drugs.

Hemodynamic Management

Because hemodynamic changes may be sudden and marked, good communication between the surgeon and the anesthesiologist is essential. It is important to understand the changes associated with the different parts of the surgery so that they can be anticipated and promptly addressed. The hemodynamic or ischemic changes that can occur are not exclusively related to any particular technique. As surgical stimulation is relatively light compared to the depth of anesthesia during this period, hypotension may ensue. These situations can usually be treated with fluid loading or with vasopressors.

Phenylephrine is the most commonly-used vasopressor; however, on rare occasions, norepinephrine or epinephrine is required. Phenylephrine is often chosen because of its predominantly α-adrenergic activity. The arterial and venous vasoconstriction produced increases blood pressure by augmenting preload and after-load. Ideally, β-agonist drugs are avoided before and during coronary revascularization because they increase myocardial and systemic oxygen consumption. In case of depressed pre-operative ventricular function, inotropes and/or pre-operative insertion of an intra-aortic balloon pump are mandatory.

Low-dose nitroglycerin infusion is normally initiated at the beginning of the procedure (0.1-0.3 µg/kg/min). By increasing sub-endocardial flow and decreasing preload, nitroglycerin can improve myocardial perfusion. It also prevents arterial graft vasospasm.[31]

Hemodynamic Management Related to Specific Territories of Revascularization

The vessel with the most severe stenosis is generally revascularized first. Unfractionated heparin 100 U/kg is administered before the anastomosis is started to prevent intra-coronary thrombus formation during vessel manipulation. Should prompt institution of CPB be necessary, heparin is added to achieve a total dose of 300 U/kg. In OPCAB cases, heparin therapy is monitored by activated clotting time (ACT) measured every 30 minutes. The arbitrary aim is ACT between 300 and 400 seconds, and small heparin boluses of 25-50 U/kg are added to maintain this range. Antifibrinolytic drugs are generally not given.

The CPB is set up but not primed. The perfusionist is on stand-by and always present in the operating room. The main reasons why a planned OPCAB is converted to an on-pump case are: hemodynamic compromise that is unresponsive to corrective measures, refractory arrhythmias, persistent ischemic changes or a deep intra-myocardial left anterior descending artery where right ventricle injury is feared.

Mechanical coronary artery stabilizers exist in two categories: those that work by compression and those that use tissue suction. These stabilizers have greatly facilitated surgery on the beating heart by maintaining a relatively stable surgical field even at heart rates of 100/min. Therefore, pharmacological manipulations are now rarely necessary to assist the surgeon. However, small doses of beta-blockers can be administered occasionally (metoprolol 2.5-5 mg IV) to maintain heart rate in the range of 50-75 /min to decrease myocardial oxygen consumption during coronary occlusion without a deleterious effect on systemic perfusion.

The left anterior descending (LAD) artery is particularly well suited for bypass because of its anterior location. It requires little mobilization for exposure and, hence, the hemodynamic disturbance is modest. Generally, the anterior wall is elevated with pericardial traction sutures or a sponge placed behind the heart. This may lead to a short period of mild, self-limited hypotension. LAD occlusion can produce local ischemia of the anterior wall, causing systolic and diastolic dysfunction. Grafting of the right coronary artery, and particularly the circumflex coronary arteries, is sometimes associated with more hemodynamic and electrical instability.[32] Despite technical advances in beating heart surgery, grafting of the circumflex coronary artery is still the most challenging part of the operation for both the surgeon and the anesthesiologist. Temporary occlusion of the proximal right coronary artery can be challenging too by causing atrioventricular node ischemia. The anesthesiologist has to be prepared to deal with bradycardia, temporary atrio-ventricular block, and hypotension. These events are difficult to predict. Displacement of the heart required for surgical exposure causes changes in the geometry of the ventricular cavities. Animal studies have shown that elevating the apex to rotate the heart upwards 90° leads to decreases in stroke volume, cardiac output and blood pressure despite increases in pressures and volumes of the right heart.[33] These are corrected by Trendelenburg positioning. Grundeman et al[34] showed that coronary flow during displacement of the heart is diminished in all coronaries but is most marked in the circumflex. Collaterals modify the hemodynamic response to surgical positioning of the heart and to coronary blood flow interruption. Koh et al[35] using data derived from the TEE and pulmonary artery catheter during off-pump surgery, found systolic dysfunction in all patients studied, but diastolic dysfunction was only present in those with little or no collateral flow in the relevant coronary artery territory.

In the current era of off-pump coronary surgery, inotropic support is required less often than in the past. Deep pericardial traction sutures placed near the left pulmonary veins and the inferior vena cava (IVC) allow the heart to be positioned vertically while maintaining near normal venous return, thus reducing hemodynamic instability. Trendelenburg positioning aids surgical exposure of the inferior and posterior walls of the heart and improves venous return.[33] It prevents hypotension during surgical manipulations. Judicious fluid loading helps to maintain adequate ventricular filling while the heart is verticalized. The risk of fluid overloading and the resultant ventricular dysfunction as the heart is returned to the pericardium must always be kept in mind. Gentle surgical manipulation and a period of physiological equilibration before starting the anastomosis permit surgical access to all territories in the vast majority of patients.

Occasionally, surgical positioning and short periods of coronary artery occlusion precipitate increased pressure in the right heart cavities (central venous pressure and pulmonary artery pressure) indicating acute decreased right ventricle compliance or temporary left ventricular failure. These changes may be correlated with an ECG suggestive of ischemia. The initial management approach is to infuse nitroglycerin at 1-3 µg/kg/min to increase venous capacitance and reduce ventricular distention. It is important to maintain coronary perfusion pressure, and phenylephrine is used for this purpose. Partial IVC snaring can also rapidly or gradually diminish the right ventricular preload. This will have immediate repercussion on left ventricular hemodynamics. In our experience with several of these cases, the maneuver was found to be very efficient in helping the left ventricle to resume normal contraction and give the anesthesiologist time for inotropic readjustment.[35] If these measures are ineffective and hemodynamic instability persists, the surgeon may consider altering surgical positioning of the heart, use of an intra-coronary shunt, insertion of an intra-aortic balloon pump, or surgical revascularization under CPB assistance.

Proximal anastomose of saphenous vein bypass grafts are performed with side-clamping of the ascending aorta. To reduce the risk of associated embolic events and to minimize the risk of aortic dissection, the ascending aorta is generally manipulated only once during the procedure.[36] Aortic wall stress is decreased by lowering systemic blood pressure to 90 mmHg systolic by either deepening the anesthetic or by administering nitroglycerin or nitroprusside during the period of side-clamping. Many authors have proposed epiaortic scanning, particularly in high-risk patients, to reduce the risk of embolic events.[37,38] Manual palpation of the aorta at the time of clamping is not a sensitive method of detecting atheromatous plaques.[39] Epi-aortic scanning and TEE are much more sensitive ways of detecting aortic disease. In the presence of marked atheromatous disease, proximal anastomoses should be performed with a technique that does not require ascending aortic clamping.[40]

When the grafts are completed, the quality of the bypass can be evaluated in several ways. A Doppler study that documents good diastolic flow normally reflects adequate revascularization.[41] A return to baseline of the ST segment and T-wave morphology is also an indirect sign of restored coronary blood flow. When the surgical result is satisfactory, anti-coagulation is reversed with protamine sulfate, given in a ratio of 10 mg per 2,000 U of the initial heparin dose.

Blood Salvaging Strategies

OPCAB surgery is associated with decreased transfusion requirements.[42] Avoiding CPB eliminates dilutional anemias related to the pump. By causing fewer inflammatory and hemostatic changes, OPCAB surgery leads to fewer coagulation abnormalities and decreased bleeding in the immediate post-operative period. A cell saver apparatus is generally used when intra-operative blood loss exceeds 400 ml. Antifibrinolytic agents are not routinely administered for OPCAB surgery.

Fluid Replacement

Continuous fluid replacement throughout surgery is necessary to maintain preloading of the heart. It is common to end up with a hypovolemic patient at completion of the procedure. This is often explained by underestimation of undetectable losses and peripheral vasodilatation secondary to the inflammatory response due to the surgical trauma. It is noteworthy to acknowledge that the undetectable losses through an open thorax can easily reach 6-8 ml/kg/hr.[43] Combined with the fasting deficit, which is about 1.5 cc/kg per hour of fasting, these fluid losses can attain significant levels. Monitoring of urine output is also mandatory to constantly assess the patient's circulating volume and renal blood flow. The colloid versus crystalloid debate is still unresolved, and so the choice of intravenous fluids is left to the anesthesiologist's preference.

Immediate Post-Operative Period

Our general policy with regard to mechanical ventilation is to aim for early extubation or the so-called "fast-track" course. The anesthetic technique should permit tracheal extubation within a few hours of completion of surgery.[44,45] Numerous studies have failed to relate any benefit to the length of post-operative mechanical ventilation.[46] On the other hand, early extubation can shorten ICU time and reduce total cost of the procedure by close to 25%.[47,48] Our practice consists of a certain period of observation and stabilization prior to extubation. The patient is weaned off mechanical ventilation only once the usual criteria have been met. These are: awareness by the patient, an unobstructed airway, hemodynamic stability with minimal inotropic requirement, low mediastinal bleeding (<100 ml/hour), and adequate gas exchange.

Post-Operative Analgesia

Early extubation and fast tracking rely on good post-operative analgesia. This is best achieved with a multimodal approach where benefits are drawn from each class of drug while minimizing dose-related side-effects. The cornerstone of pain relief is the opioid class of drugs. Morphine (2-4 mg/hour) and fentanyl (25-75 μg/hour) are commonly administered intravenously, either as an infusion or via a patient-controlled analgesia device. As oral medication is tolerated, hydromorphone or acetominophen-codeine combination is given. Nonsteroidal anti-inflammatory drugs are good coanalgesics with a significant opioid-sparing action.[49] They are often omitted in the immediate post-operative period (first 24 hours) because of possible side-effects: platelet dysfunction, renal impairment and gastrointestinal tract bleeding. Anti-COX2 drugs with their longer duration of action and promising safety profile are a welcome addition.

Neuraxial analgesia, either spinal or epidural, is another approach to post-operative pain management. The advantages of profound analgesia with little respiratory depression, attenuation of the stress response and decreased sympathetic tone are appealing. Hynninen et al[50] reported that 5 μg/kg of intrathecal morphine produced excellent analgesia, improved pulmonary function tests, and led to better cognitive function. Careful patient selection, performance of the technique the day before surgery, and anti-coagulation may explain their good results. This practice is, however, logistically difficult. We rarely use epidural or spinal analgesia in our institution. The benefits of the technique must be weighed against the rare but serious complication of epidural hematoma and spinal cord compromise. This complication is fortunately rare with an incidence of 1 in 200,000. The cardiac surgical group may be at higher risk because of pre-operative anti-coagulation or exposure to other drugs that have an effect on hemostasis.

Conclusion

The anesthetic management of patients undergoing an OPCAB procedure is similar to conventional CABG. However, sudden hemodynamic changes can occur due to surgical manipulation of the heart and iatrogenic ischemia. These changes must be anticipated and managed carefully to permit the successful completion of surgery and complete revascularization without the need for CPB. The importance of good communication between the surgeon and anesthesiologist cannot be over-emphasized.

References

1. Ascione R, Lloyd CT, Underwood MJ et al. Inflammatory response after coronary revascularization with or without CPB. Ann Thorac Surg 2000; 69:1198-204.
2. Brasil LA, Gomes WJ, Salomao R. Inflammatory response after myocardial revascularization with or without cardiopulmonary bypass. Ann Thorac Surg 1998; 66:56-9.

3. Murkin JM, Boyd WD, Ganapathy S. Beating heart surgery: Why expect less central nervous system morbidity? Ann Thorac Surg 1999; 68:1498-501.

4. Boyd WD, Desai ND, Del Rizzo DF. Off-pump surgery decreases postoperative complications and resource utilization in the elderly. Ann Thorac Surg 1999; 68:1490-93.

5. Tremblay NA, Hardy JF, Perreault J. A simple classification of the risk in cardiac surgery: The first decade. Can J Anesth 1993; 40:103-11.

6. Murkin JM, Boyd WD, Ganapathy S et al. Postoperative cognitive dysfunction is significantly less after coronary artery revascularization without cardiopulmonary bypass. Ann Thorac Surg 1999; 68:1469(Abstract).

7. Brabant SM, Bertrand M, Eyraud D et al. The hemodynamic effects of anesthetic induction in vascular surgical patients chronically treated with angiotensin II receptor antagonists. Anesth Analg 1999; 89:1388-92.

8. Licker M, Schweizer A, Hohn L et al. Cardiovascular responses to anesthetic induction in patients chronically treated with angiotensin-converting enzyme inhibitors. Can J Anesth 2000; 47:433-40.

9. Rackow, EC. Pulmonary artery catheter consensus conference; consensus statement. Crit Care Med 1997; 25(6):901.

10. CAS Guidelines to the practice of anesthesia 2000. Can J Anesth 2000; 47(suppl).

11. London MJ, Hollenberg M, Wong MG et al. Intraoperative myocardial ischemia: Localization by continuous 12-lead electrocardiography. Anesthesiology 1988; 69:232-41.

12. ASA/SCA Practice guidelines for perioperative transesophageal echocardiography. Anesthesiology 1996; 84:986.

13. Couture P, Denault A, McKenty S et al. Impact of routine use of intraoperative transesophageal echocardiography during cardiac surgery. Can J Anesth 2000; 47:20-6.

14. Katz ES, Tunick PA, Rusinek H et al. Protruding aortic atheromas predict stroke in elderly patients undergoing cardiopulmonary bypass: Experience with intraoperative transesophageal echocardiography. J Am Coll Cardiol 1992; 20:70-7.

15. Dennehy KC, Nathan HJ. The effect of mild hypothermia on bleeding following coronary artery bypass graft surgery. Anesth Analg 1997; 84:570.

16. Kurz A, Sessler DJ, Lenhardt R. Perioperative normothermia to reduce the incidence of surgical wound infection and shorten hospitalization. Study of wound infection and temperature group. N Engl J Med 1996; 334:1209-15.

17. Frank SM, Beattie C, Christopherson R et al. Unintentional hypothermia is associated with postoperative myocardial ischemia. The Perioperative Ischemia Randomized Anesthesia Trial Study Group. Anesthesiology 1993; 78:468-76.

18. Howie MB, Cheng D, Newman MF et al. A randomized double-blinded multicenter comparison of remifentanil vs fentanyl when combined with isoflurane/propofol for early extubation in coronary artery bypass graft surgery. Anesth Analg 2001; 92:1084-93.

19. Lowenstein E, Hallowell P, Levine F et al. Cardiovascular response to large doses of intravenous morphine in man. N Engl J Med 1969; 281:1389-92.

20. Stanley TH, Webster LR. Anesthetic requirements and cardiovascular effects of fentanyl-oxygen and fentanyl-diazepam-anesthesia in man. Anesth Analg 1978; 57:411-16.

21. Hanaleshka A, Jacobsohn E. Ischemic preconditioning: Mechanisms and potential clinical applications. Can J Anesth 1998; 45:670-82.

22. Galinanes M, Argano V, Hearse DJ. Can ischemic preconditioning ensure optimal myocardial protection when delivery of cardioplegia is impaired? Circulation 1995; 92(suppl II):II-389.

23. Jacobsohn E, Young CJ, Aronson S et al. The role of ischemic preconditioning during minimally invasive coronary artery bypass surgery. J Cardiothorac Vasc Anesth 1997; 11:787-92.

24. Kersten JR, Schmeling TJ, Hettrick DA et al. Mechanism of myocardial protection by isoflurane. Role of adenosine triphosphate-regulated potassium (Katp). Anesthesiology 1996; 85:794-807.

25. Cason BA, Gamperl AK, Slocum PE et al. Anesthetic-induced preconditioning: Previous administration of isoflurane decreases myocardial infarct size in rabbits. Anesthesiology 1997; 87:1182-90.

26. Belhomme D, Reynet J, Louzy M et al. Evidence for preconditioning by isoflurane in coronary artery bypass graft surgery. Circulation 1999; 100(suppl II):II-340.

27. Maslow AD, Regan MM, Heindle S et al. Postoperative atrial tachyarrhythmias in patients undergoing coronary artery bypass surgery without cardiopulmonary bypass: A role for intraoperative magnesium supplementation. J Cardiothorac Vasc Anesth 2000; 14:524-30.

28. Speziale G, Ruvolo G, Fattouch K et al. Arrhythmias prophylaxis after coronary artery bypass grafting: Regiments of magnesium sulfate administration. J Thorac Cardiovasc Surg 2000; 48:22-6.

29. Bert AA, Reinert SE, Singh AK. A beta-blocker, not magnesium, is effective prophylaxis for atrial tachyarrhythmias after coronary artery bypass graft surgery. J Cardiothorac Vasc Anesth 2001; 15:204-9.

30. Teragawa H, Kato M, Yamagata T et al. The preventive effect of magnesium on coronary spasm in patients with vasospastic angina. Chest 2000; 118:1690-5.

31. Shapira OM, Alkon JD, Macron DS et al. Nitroglycerin is preferable to diltiazem for prevention of coronary bypass conduit spasm. Ann Thorac Surg 2000; 70:883-8.

32. Do QB, Chavanon O, Couture P et al. Hemodynamic repercussion during beating heart CABG surgery. Can J Cardiol 1999; 15(suppl D):177D.

33. rundeman PF, Borst C, van Herwaarden JA et al. Vertical displacement of the beating heart by the octopus tissue stabilizer: Influence on coronary flow. Ann Thorac Surg 1998; 65:1348-52.

34. Grundeman PF, Borst C, Van Herwaarden JA et al. Vertical displacement of the beating heart by the octupus tissue stabilizer: Influence on coronary flow. Ann Thorac Surg 1998; 65:1348-52.

35. Koh TW, White GS, DeSouza AC et al. Effect of coronary occlusion on left ventricular function with and without collateral supply during beating heart coronary artery surgery. Heart 1999; 81:285-91.

36. Dagenais F, Cartier R. Pulmonary hypertension during beating heart coronary surgery: Intermittent vena cava snaring. Ann Thorac Surg 1999; 68:1094-5.

37. Chavanon O, Carrier M, Cartier R et al. Increased incidence of acute ascending aortic dissection with off-pump aortocoronary bypass surgery? Ann Thorac Surg 2000; 71:117-21.

38. Stump DA, Newman SP. Embolic detection during cardiopulmonary bypass. In: Tegler CH, Babikian VL, Gomez CR, eds. Neurosonology. St. Louis, Mosby: 1996:252-55.

39. Murkin JM. Epiaortic scanning decreased cerebral emboli during aortic canulation and application of partial occlusion clamp. Heart Su:g Forum 2003; 6(4):203-4.

40. Wareing TH, Davila-Roman VG, Barzilai B et al. Management of the severely atherosclerotic ascending aorta during cardiac operations. J Thorac Cardiovasc Surg 1992; 103:453-62.

41. Trehan N, Mishra M. Surgical strategies in patients at high risk for stroke undergoing coronary artery bypass grafting. Ann Thorac Surg 2000; 70:1037-45.

42. Dagenais F, Perrault LP, Cartier R et al. Beating heart coronary artery bypass grafting: Technical aspects and results in 200 patients. Can J Cardiol 1999; 15:867-72.

43. Nader ND, Khadra WZ. Blood product use in cardiac revascularization: Comparison of on- and off-pump techniques. Ann Thorac Surg 1999; 68:1640-3.

44. Cartier R, Semper Dias O, Pellerin M et al. Changing flow pattern of the internal thoracic artery undergoing coronary bypass grafting: Continuous-wave Doppler assessment. J Thorac Cardiovasc Surg 1996; 112:52-8.

45. Prakash O, Jonson B, Meij S et al. Criteria for early extubation after intracardiac surgery in adults. Anesth Analg 1977; 56:703-8.

46. Quasha AL, Loeber N, Feeley TW et al. Postoperative respiratory care: A controlled trial of early and late extubation following coronary artery bypass grafting. Anesthesiology 1980; 52:135-41.

47. Nicholson DJ, Kowalski SE, Hamilton GA et al. Postoperative pulmonary function in CABG patients undergoing early tracheal extubation: A comparison between short-term mechanical ventilation and immediate extubation. J Cardiothorac Vasc Anesthesia 2002; 16:27-31.

48. Cheng DCH, Karski J, Peniston C et al. Morbidity outcome in early versus conventional tracheal extubation after coronary artery bypass grafting: A prospective randomized controlled trial. Anesthesiology 1996; 85:1300-10.

49. Eschun GM, Jacobsohn E, Roberts D et al. Ethical and practical considerations of withdrawal of treatment in the intensive care unit. Can J Anaesth 1999; 46:497-504.

50. Hynninen MS, Cheng DC, Hossain I et al. Nonsteroidal anti-inflammatory drugs in the treatment of postoperative pain after cardiac surgery. Can J Anesth 2000; 47:1182-7.

CHAPTER 6

Indications and Surgical Strategies for OPCAB

Nicolas Dürrleman and Raymond Cartier

Introduction

Since its reintroduction in the late 1990s, OPCAB surgery has been adopted by the international community in a proportion fluctuating from 1% to 98%.[1,2] This large variability clearly reflects an incredible disparity in acceptance of the procedure. Diegler and his team from Leipzig initially proposed a series of indications regarding the procedure.[3] With time, others encouraged a more liberal use of off-pump surgery.[4] Ultimately, it became apparent that the limiting factor remained the surgeon's experience. The main indication for going off-pump is obviously to decrease the inherent morbidity related to the use of the cardiopulmonary bypass (CPB) especially among patients with significant co-morbid factors. Other indications are clotting disturbance, a calcified ascending aorta, and minimization of blood transfusions. But down the road, the final decision depends on the coronary anatomy, the shape of the myocardium, and tolerance of the patient's hemodynamics to surgical mobilizations and manipulations. Table 1 depicts the major and relative contraindications to OPCAB surgery that we have applied in our 7-year experience over 900 cases. The purpose of this chapter is to review the different surgical strategies that we have developed and adapted to optimize the OPCAB procedure and propose a "survival guide" to OPCAB surgery.

Approaching the Left Anterior Descending Artery (LAD) and Diagonal Artery

The LAD artery is the easiest target to reach. Being located on the anterior face of the heart, minimal surgical mobilization is required. Although a simple sponge placed behind the heart is necessary to move the apex in the surgical field, we have opted through the years for a "single" stitch approach (Fig. 1). The free edge of the left side of the pericardium is held and lifted up by the assistant as the operator anchors a single stitch just above the phrenic nerve where the apex of the heart normally bulges in the pericardium. This brings the apex of the myocardium straight on the mid-line and has the advantage of being stable throughout the suturing period. Classically, sponges placed behind the heart have the disadvantage of gradually getting wet, then losing volume and necessitating repositioning by the surgeon. The diagonal artery (DA) is generally approached in a similar fashion. A second stitch close to the left inferior pulmonary vein (LIPV) is normally needed to create more myocardial rotation and facilitate lateral access. The table is generally rotated 20 degrees towards the surgeon with slight Trendelenburg positioning.

Table 1. Contraindications to OPCAB surgery

Absolute:
- Pre-operative hemodynamic instability
- Deep myocardial Left anterior descending artery
- Moderate (3+) or severe (4+) mitral insufficiency

Relative:
- Pulmonary hypertension
- Diffuse coronary artery disease
- Dense myocardial adhesion during reoperative surgery
- Enlarged ascending aorta
- Left main disease and a concurrent non-reconstructable coronary network

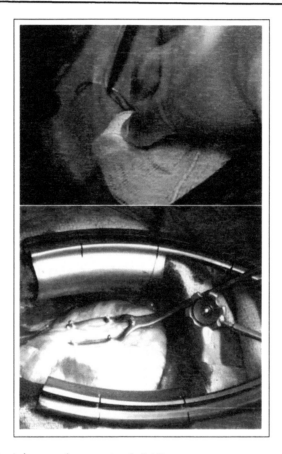

Figure 1. The single stitch approach to exposing the LAD artery.

Special care should be taken with positioning of the stabilizer on the DA (as mentioned in Chapter 7). The DA crosses the path of the left ventricular outflow tract (LVO). Any significant deformation of the funnel shape of the LVO tract will create obstruction and build up left ventricular afterload (Fig. 2). The LAD artery running on top of the septum has a less compromising effect during stabilization.

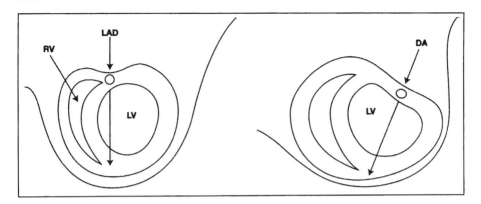

Figure 2. Schematic showing the potentially deleterious effect of outflow tract compression of the left ventricle (LV) during stabilization of the diagonal artery (DA) compared to the LAD artery where compression occurs mainly on the septum. RV: right ventricle.

Approaching the Posterior Descending Artery (PDA)

The PDA is approached in like manner. The apex is brought on the mid-line, as described with the LAD artery. The table is set in a 30-degree "head down" position with a 30-degree tilt towards the right-sited operator. The closest to the base of the heart the target vessel is the easier the coronary stabilization will be. When the mid-PDA is targeted for anatomical reasons, oscillation of the apex frequently impedes satisfactory stabilization. Apical suction in these circumstances enhances the stabilization (Fig. 3).

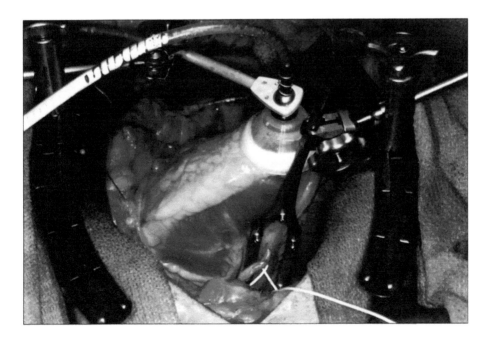

Figure 3. View of the inferior wall during grafting of the mid-PDA. See text for explanation.

Approaching the Right Coronary Artery (RCA)

As with the LAD artery, the RCA is one of the target vessels that requires the least surgical manipulations. To bring the RCA in the middle of the surgical field, 2 pericardial stitches (Fig. 4) are anchored, one just above the inferior vena cava (IVC), and the other in the same plane just above the left atrium. The table is rotated 20 degrees opposite to the operator with 20–30-degree head-down tilt positioning. This brings the mid-RCA straight on the midline and really eases positioning of the stabilizer.

Approaching the Obtuse Marginal Artery (OMA)

Exposure of the circumflex territory has been described before.[5] We developed this approach in 1996, which consisted of 4 deep pericardial stitches ranging between the left superior pulmonary vein (LSPV) and the IVC. The technique is quite comparable to the "Lima stitches" introduced in North America in 1997 by Tom Salerno.[6] Although some warning[7] has been voiced about the risk of retro-mediastinal bleeding with these deep pericardial stitches, we have not, in our experience, encountered severe bleeding from pericardial stitch insertion. We do express some concern though about how deep these stitches should be inserted. The lungs should be deflated first. Then, the assistant has to pull the free edge of the pericardium while his right hand, holding the wall suction, retracts the pericardium. This generously enhances the vision field of the operator. Obviously, care has to be taken not to anchor these stitches too deeply, especially toward the pleural space. We have observed once in a while, persistent left-sited pneumothorax. We now routinely leave the left chest tube for a minimum of 48 hours, generally removing it after this period. The 4 stitches are superficial and located close to the base of the heart at about 1-2 cm away from the pericardial attachment to the heart. The surgeon has to pass the needle of the pericardial suture away from the heart and not towards it to avoid myocardial injury. The first suture is anchored just above the LSPV, the second below the LIPV, the third one called, "the intermediate", is located between the LPIV vein and the IVC, and the fourth one is very close to the IVC (Fig. 5). Once completed, this generally provides a complete verticalization of the heart. Then, the surgeon can position the stabilizer with minimal myocardial contact, minimizing further hemodynamic disturbances. Because all the coronary arteries converge towards the apex, by controlling the apex, the operator theoretically has access to all of them (Fig. 6).

Coronary Artery Anastomosis

The suturing technique that we use is similar to what we were familiar with in conventional on-pump surgery. This generally consists of "parachuting" 3 or 4 suspended stitches before completing the running suture. Polypropylene 7-0 or 8-0 sutures are deployed for internal thoracic artery (ITA) grafting (sequential and distal), according to the size of the distal vessels. A bloodless field is obtained by gently snaring the target coronary artery with silicone vessel loops. Proximally, the loops are crossed over, whereas distally, a single loop is used to avoid damaging the distal bed (Fig. 7). Recently developed silicone pledgets obviate for the need of loop-crossing. It favors a rather more "antero-posterior compression" than a "circumferential constriction" during vessel occlusion with lesser traumatic consequences (Fig. 7B). Tension is kept minimal distally by fixing it directly on the stabilizer (CorVasc, CoroNéo, Montreal, PQ, Canada), which allows better tension control. Lately, some authors have advocated the systematic use of shunts during the grafting period.[8] Numerous publications have expressed concerns about the effect of shunts on coronary artery endothelial function.[9-11] Although it has been demonstrated experimentally that endothelial damage can be attenuated by decreasing the size of the shunt, endothelial injury still persists.[10] At the opposite end, silicone loops were shown to create minimal injury to the endothelium compared to polypropylene snaring, and endovascular shunt (cf. Chapter 13).[12] We do employ shunts only in specific circumstances, which will be discussed later. An important

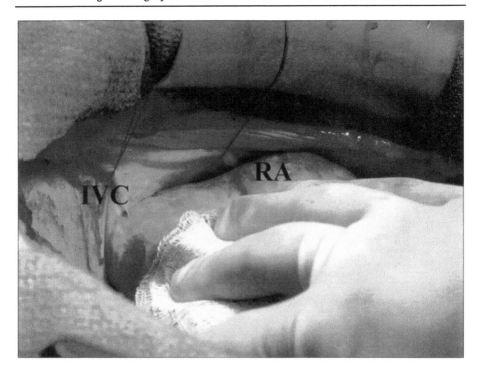

Figure 4. View of the inferior vena cava (IVC), the right atrium (RA), and the pericardial stitches used to expose the distal right coronary artery.

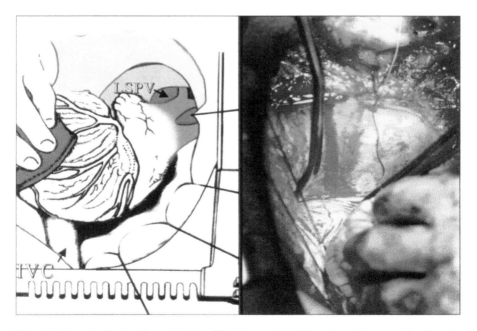

Figure 5. Schematic showing the positioning of the 4 deep pericardial traction stitches spreading from the left superior pulmonary vein (LSPV) to the inferior vena cava (IVC).

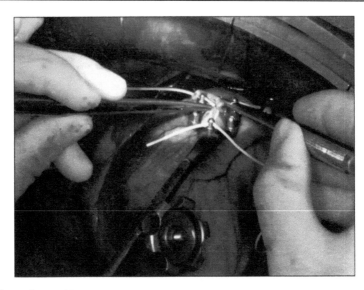

Figure 6. Surgeon's view of the posterior wall during coronary grafting once verticalization has been completed.

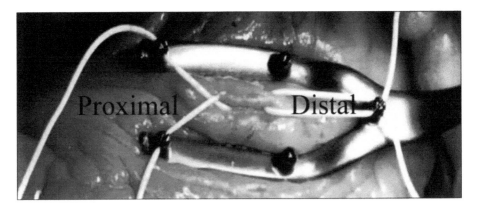

Figure 7A. Silicone loop snaring. Distally, a single loop is used to decrease potential intimal injury.

aspect that is worth mentioning is needle positioning during graft suturing. To minimize the risk of capturing the posterior wall of the grafted coronary artery, the tip of the needle should always be kept vertical during its introduction in the native coronary artery. The mid-portion of the needle is maintained parallel to the back wall of the vessel. In this setting, the operator can position every single stitch quite precisely. Once the graft anastomosis is completed, its patency is verified with Doppler probe (Smartdp, Hadeco, 8 MHz). On the left coronary network, we pay, a special attention to diastolic velocity (> 0.8 KHz) and the diastolic to systolic velocity ratio (2 or more). We do not hesitate to refashion the anastomosis if there is doubt about graft patency (up to 8-10% in our practice). On the right coronary network, we look for equalization of the systolic-diastolic ratio; the RCA normally gets half of its blood flow during systole, which explains the different pattern of the signal.

Figure 7B. Silicon pledget and lopp for soft and atraumatic proximal and distal vessel occlusion ("Swift loop", CoronoNéo, Montreal, PQ, Canada).

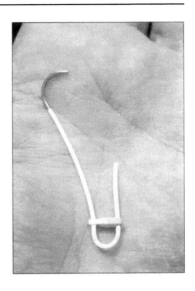

Surgical Strategies (Table 2)

One of the goals sought by surgeons during OPCAB surgery is the avoidance of conversion to CPB. To minimize ischemia during surgical manipulation and the risk of destabilizing the patient's hemodynamic we have established and followed, early in our experience, a certain number of rules. First, the most stenotic vessel is always revascularized first. This vessel is normally well collateralized and supports occlusion without life-threatening myocardial ischemia. Second, when a saphenous vein or a radial artery conduit is used, the proximal anastomosis is completed once the distal anastomosis is completed, before the next distal anastomosis. Ante-grade flow provides needed supportive blood flow through the collaterals during the next coronary cross-clamping. Normally the RCA or the LAD artery is revascularized first. Third, we generally perform a pre-ischemic test of 2 minutes before grafting of the LAD artery, especially when the stenosis is not critical. Shunts are employed only when ischemia is suspected. Fourth, the ascending aorta is side-clamped only once during the procedure to minimize aortic trauma and all the proximal anastomosis are completed during the same period. Fifth, systemic pressure is always reduced around 90-100 mm Hg before and during the entire side-clamping of the ascending aorta. Sixth, the anterior and the inferior territories are always revascularized before the posterior territory.

Table 2. Surgical strategies during OPCAB surgery

1. "Culprit lesion first"
2. Antegrade flow established before second anastomosis
3. Pre-ischemic test prior to LAD artery grafting
4. Single side-clamping of the ascending aorta
5. Moderate hypotension during proximal anastomosis
6. Revascularization of the anterior and inferior territories prior to the circumflex artery

---

Special Situations

Dealing with the "Big Heart"

Dilated ischemic cardiomyopathic hearts are, without any doubt, the most difficult cases to expose and stabilize during OPCAB surgery. In these circumstances, the working space between the posterior wall of the heart and the left arm of the retractor is reduced during verticalization. To increase this working space, the surgeon has the choice between a large opening of the right pleura or a separation of the diaphragmatic and pleural pericardium. Through the years, we have adopted the latter. This "pleuro-diaphragmatic disconnection" is normally carried out with the electro-cauterizer at moderate power to avoid injuring the phrenic nerve. It could be further extended toward the RIPV (Fig. 8). This will deepen the base of the heart and facilitate the work of the surgeon by opening up the working angle on the posterior wall (Fig. 9 A-B).

Reoperative Surgery

A frequent situation in reoperative surgery is the presence of an occluded native coronary artery whose vascular supply depends of a stenotic graft with impaired blood flow (Fig. 10). Frequently, the atheromatous graft is the culprit lesion and is responsible for the recurrent angina. Manipulations of this graft have to be undertaken with a great deal of gentleness to avoid creating any further ischemic hemodynamic instability. The optimal strategy in these situations is to use the ITA as proximal inflow when available. This avoids side-clamping of the ascending aorta and further exacerbation of ischemia. If only 1 ITA is available, or if no ITA is available at all, the surgeon has to plan out in advance end-to-side reconstruction with either a radial artery conduit or a saphenous vein segment branching off the ITA conduit, if available, or the brachiocephalic artery, if not.[13] The subclavian and radial artery network should preferably be scanned prior to surgery with special attention being paid to the patency of the brachiocephalic artery, which could serve as a potential site for proximal anastomosis. Other alternatives are the new aortic connectors that allow proximal anastomosis without side- clamping (which will be discussed in another chapter).

The Calcified Ascending Aorta

With aging of the surgical population, assessment of the ascending aorta prior to side-clamping has become a necessity. Lately, we have routinely adopted, in patients aged 75 years and older, epiaortic scanning of the ascending aorta (Fig. 11). Many authors have reported on the efficiency of this tool for detection of aortic arteriosclerosis.[14,15] This has led us to occasionally change our strategy of revascularization, mainly with the more aggressive "no-touch" technique, that also appears to be beneficial for improved neurological outcome.[16,17]

The Sudden Pulmonary Artery Pressure Rise

Occasionally, due to surgical manipulations, or more frequently to myocardial ischemia, pulmonary artery pressure suddenly rises above normal during the procedure. This is generally followed by left ventricular distention and unstable hemodynamics. It is a potentially life-threatening condition that is frequently related to sudden ischemic mitral insufficiency.[18] To re-establish the patient's hemodynamics and avoid acute conversion to CPB, the IVC is temporarily cross-clamped or progressively snared.[19] By decreasing the inflow, the left ventricle shrinks rapidly and resumes normal contractility. IVC-snaring is easily tolerated for a few minutes before the occlusion is relieved progressively. The effect of this maneuver is comparable to an intravenous nitro-glycerin bolus. We currently use it in close to 10% of all our cases with no outcome side effect.[20] It gives the anesthesiologist time to readjust the pharmacological support.

Figure 8. The "pleuro-diaphragmatic disconnection" to allow herniation of the heart in the pleural space during verticalization.

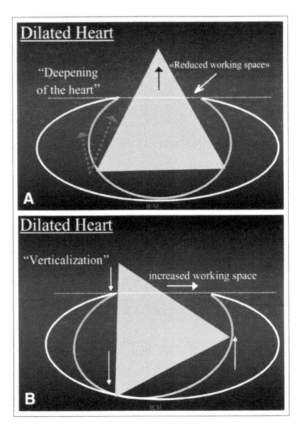

Figure 9. A) The "pleuro-diaphragmatic disconnection" used to increase working space. B) Deepening of the base of the heart increases working space on the posterior wall.

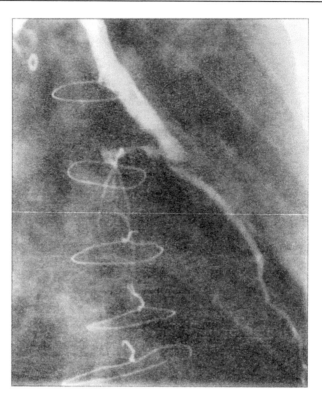

Figure 10. Frame taken from a coronary angiogram showing a highly stenotic vein graft revascularizing an occluded native LAD artery. See text for explanation.

Blood-Saving Strategy

One of the major advantages of OPCAB surgery has been the reduction of blood loss and transfusion needs during and after coronary artery revascularization. This has been reported by a majority of authors.[21] In our experience, avoiding CPB has cut down the incidence of blood product transfusion by more than half.[20] This has been made possible also by strict guidelines on blood collection and transfusion. Currently, we recycle the blood lost in every case. When blood losses remain below 400 ml, the blood is filtered and returned straight back to the patient (Fig. 12A). When it goes beyond 400 ml, the blood is processed through the cell-saver apparatus and re-infused (Fig. 12B). During the post-operative period, we also partially recover the bloodshed, but limit the amount of re-infused blood to 5-7 ml/Kg. Patients are transfused to maintain their hemoglobin count equal or above 75 gr/L.

Use of Aprotinin

The efficacy of aprotinin in reducing blood loss during conventional coronary artery surgery has been well documented.[22] A recent randomized trial has further confirmed its efficacy in OPCAB surgery.[23] Others, however, have reported early graft thrombosis with the drug and have pounded words of caution on its use.[24] In our series, aprotinin was administered to 32 OPCAB patients in the last 3 years. These were patients on antiplatelet drugs prior to surgery in whom more than the usual blood loss was anticipated. No operative death was reported. Only 1 patient sustained a delayed stroke that occurred after undergoing a femoral angioplasty following intra-aortic balloon pump withdrawal. None sustained a peri-operative myocardial

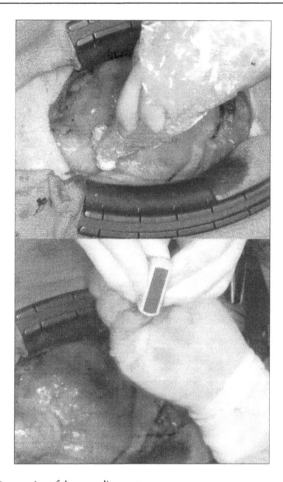

Figure 11. Epiaortic scanning of the ascending aorta.

infarction, and at 30 days there were no venous thromboembolic complications. The prevalence of transfusion was 25%. Although such a short series does not allow any strong conclusions to be drawn on the blood-saving effect of aprotinin, it substantiates its safety at short-term (up to 30 days) follow-up. All patients receive a full dose of heparin during surgery (3 mg/kg) and are kept on subcutaneous heparin (5,000 units TID) during hospitalization, ASA 80 mg (life time), and clopidogrel 75 mg (first month). Although an optimal antiplatelet regimen with OPCAB surgery still needs to be established, it is somehow important to be more aggressive than with conventional surgery. A post-operative hypercoagulable state is postulated to be present following OPCAB surgery, and could be further amplified by the use of aprotinin.[25,26] We do not administered aprotinin routinely but only in patients with clotting disturbances, in whom excessive bleeding is expected.

Conclusion

This was a brief overview of the technique we developed over the last 7 years that led us to achieve systematic off-pump surgery in close to 98% of our caseload. Undoubtedly, each surgeon has to develop his/her own approach and be comfortable with it. Early on, we opted for re-usable stabilizers for economic reasons. We have remained faithful to the same technique of

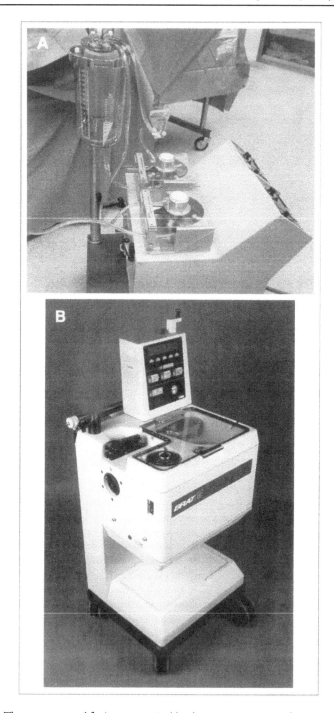

Figure 12. A) The apparatus used for intra-operative blood recovery consisting of two rotating heads, one for blood collection and one for recirculation of the stored blood. The system is primed with 1,000 units of heparin at the beginning. B) The cell-saver apparatus is used when operative blood loss exceeds 400 ml.

myocardial verticalization through all these years because of its simplicity, reliability, low cost, and reproducibility. We learned how to deal with unexpected situations and minimize the risk of precipitous conversion that could sometimes be life-threatening.[27,28] Our rate of conversion has remained low (less than 0.5%) and the rate of complete revascularization is satisfactory (> 90%). The tips and tricks that we have outlined could be a "survival guide" by any surgeon not familiar with the technique but anxious to learn it.

References

1. Aldea GS, Goss JR, Boyle EM Jr et al. Use of off-pump and on-pump CABG strategies in current clinical practice: the Clinical Outcomes Assessment Program of the state of Washington. J Card Surg. 2003; 18(3):206-15; discussion 216.

2. Cartier R, Brann S, Dagenais F et al. Systematic off-pump coronary artery revascularization in multivessel disease: experience of 300 cases. J Thorac Cardiovasc Surg 2000; 119:221-29.

3. Diegler A, Matin M, Falk V et al. Indication and patient selection in minimally invasive and off-pump' coronary artery bypass grafting. Eur J Cardiothorac Surg 1999; 16(Suppl1):S79-82.

4. Benetti F, Patel AN, Hamman B. Indications for off pump coronary surgery. J Cardiovasc Surg (Torino) 2003; 44(3):319-22.

5. Cartier R, Blain R. Off-pump revascularization of the circumflex artery: technical aspect and short-term results. Ann Thorac Surg 1999; 68:94-9.

6. Bergsland J, Karamanoukian HL, Soltoski PR et al. "Single suture" for circumflex exposure in off-pump coronary artery bypass grafting. Ann Thorac Surg 1999; 68:1428-30.

7. Salerno TA. To the Editor: A word of caution on deep pericardial sutures for off-pump coronary surgery bypass procedures. Ann Thorac Surg 2003; 76:339.

8. Balkhy HH, Quinn CC, Lois KH et al. Routine intracoronary shunting in multivessel off-pump coronary artery bypass: a retrospective review of in-hospital outcomes in 550 consecutive cases. Heart Surg Forum 2003; 6(2):E32-5.

9. Perrault LP, Menasche P, Bidouard J-P et al. Snaring of the target vessel in less invasive bypass operations does not cause endothelial dysfunction. Ann Thorac Surg1997; 63:751-5.

10. Demaria RG, Fortier S, Malo O et al. Influence of intracoronary shunt size on coronary endothelial function during off-pump coronary artery bypass. Heart Surg Forum 2003; 6(3):16-8.

11. Hangler HB, Pfaller K, Hoefer D et al. Effects of intracoronary shunts on coronary endothelial coating in the human beating heart. Heart Surg Forum 2003; 6(Suppl1):S46.

12. Okazaki Y, Takarabe K, Murayama J et al. Coronary endothelial damage during off-pump CABG related to coronary-clamping and gas insufflation. Eur J Cardiothorac Surg 2001; 19(6):834-9.

13. Tector AJ, Kress DC, Downey FX et al. Complete revascularization with internal thoracic artery grafts. Semin Thorac Cardiovasc Surg 1996;8(1):29-41.

14. Davila-Roman VG, Phillips KJ, Daily BB et al. Intraoperative transesophageal echocardiography and epiaortic ultrasound for assessment of atherosclerosis of the thoracic aorta. J Am Coll Cardiol 1996; 28(4):942-7.

15. Bonatti J. Ascending aortic atherosclerosis-a complex and challenging problem for the cardiac surgeon. Heart Surg Forum 1999; 2(2):125-35.

16. Kobayashi J, Sasako Y, Bando K et al. Multiple off-pump coronary revascularization with "aorta no-touch" technique using composite and sequential methods. Heart Surg Forum 2002; 5(2):114-8.

17. Leacche M, Carrier M, Bouchard D et al. Improving neurological outcome in off-pump surgery. The "no-touch" technique. Heart Surg Forum 2003; 6(3):169-75.

18. Couture P, Denault AY, Sheridan P et al. Partial inferior vena cava snaring to control ischemic left ventricular dysfunction [Constriction partielle de la veine cave inferieure pour controler une dysfonction ventriculaire gauche]. Can J Anaesth 2003; 50(4):404-10.

19. Dagenais F, Cartier R. Pulmonary hypertension during beating heart coronary surgery: intermittent inferior vena cava snaring. Ann Thorac Surg 1999; 68(3):1094-5.

20. Cartier R. Current trends and technique in OPCAB surgery (review). J Card Surg 2003; 18(1):32-46.

21. Ascione R, Williams S, Lloyd CT et al. Reduced postoperative blood loss and transfusion requirement after beating-heart coronary operations: a prospective randomized study. J Thorac Cardiovasc Surg 2001; 121(4):689-96.

22. Rich JB. The efficacy and safety of aprotinin use in cardiac surgery. Ann Thorac Surg 1998; 66:S6-11.
23. Engleberger L, Markart P, Eckstein FF et al. Aprotinin reduces blood loss in off-pump coronary artery bypass (OPCAB) surgery. Eur J Cardiothorac Surg 2002; 22:545-51.
24. Donias HW, Karamanoukian RL, Karamanoukian HL. Antifibrinolytic therapy during OPCAB surgery: a word of caution. J Cardiothorac Vasc Anesth 2002; 16(3):391-2.
25. Quigley RL, Fried DW, Pym J et al. Off-pump coronary artery bypass surgery may produce a hypercoagulable patient. Heart Surg Forum 2003; 6(2):94-8.
26. Feindt P, Seyfert U, Huwer H et al. Is there a phase of hypercoagulability when aprotinin is used in cardiac surgery? Eur J Cardiothorac Surg1994; 8:308-14.
27. Soltoski P, Salerno T, Levinsky L et al. Conversion to cardiopulmonary bypass in off-pump coronary artery bypass grafting: its effect on outcome. J Card Surg 1998; 13(5):328-34.
28. Mujanovic E, Kabil E, Hadziselimovic M et al. Conversions in off-pump coronary surgery. Heart Surg Forum 2003; 6(3):135-7.

Effects of Vascular-Interrupting and Hemostatic Devices on Coronary Artery Endothelial Function in Beating Heart Coronary Artery Bypass Surgery

Roland G. Demaria and Louis P. Perrault

Introduction

Coronary artery bypass grafting was first conceived and experimented on by the French Nobel Prize winner in Medicine Alexis Carrel at the beginning of the previous century.[1] Sabiston, in 1962, performed the first aortocoronary venous bypass graft in humans, and Kolesov, the first left internal mammary artery (IMA) to left anterior descending (LAD) coronary artery in 1966.[2,3] All these operations were undertaken on the beating heart. At the end of the 1960s, Favaloro and the Cleveland Clinic team launched the era of modern coronary artery bypass surgery with the use of the extra-corporeal circulation.[4] All these pioneers were confronted by the problem of blood flow control at the anastomotic site. Different techniques, such as compression, irrigation of the area or external cross-clamping with poor stabilization were tried. Rapidly, cardiopulmonary bypass (CPB) was almost universally adopted for coronary bypass surgery, allowing surgeons to achieve a bloodless and motionless operative field. The majority of coronary operations were soon performed with this technique, and beating heart coronary revascularization was abandoned as a routine procedure. Because CPB caused a major systemic inflammatory response with subsequent risks of hemorrhagic and neurological complications, and perhaps for economic reasons, some surgical teams remained faithful to the technique and were involved in its revival. In the last decade, beating heart coronary artery bypass surgery has regained tremendous popularity in the cardiovascular community.[5,6] Nevertheless, specific technical difficulties are still associated with this approach; heart stabilization or coronary bleeding at the anastomotic site. These issues may alter the quality of anastomosis and result in a greater need for reoperation or percutaneous coronary angioplasty.[7] Therefore, the ideal technique to obtain a bloodless field on coronary anastomotic sites, mandatory for the surgeon's optimal visualization, remains unresolved.[8]

Various systems were developed in the past and used for a long time to create blood flow interruption. Among them, direct artery clamping, intra-coronary shunting, different kinds of snaring sutures, blowers, gas jet, and extraluminal occlusive balloon have been tested.[9-12] Concomitantly to these new technical approaches, some fundamental observations were made on the biology of the arterial wall, especially on the role of the endothelium in vascular homeostasis.[13-17] The basic function of endothelial cells in the regulation of vascular tone and the control of coagulation has been established indubitably. Preservation of the structural and functional integrity of the endothelial layer was thus recognized as being essential in vascular reconstruction.[13,17,18] Knowledge of the biological impact of coronary occlusion or of hemostasis with different devices has become important.

The aim of this review is to assess the effects of different hemostatic devices, used in beating heart coronary artery bypass surgery, on the endothelial structure and function of coronary arteries and grafts.

Hemostatic Devices and the Endothelium

External Clamping

Direct vascular clamping upstream and downstream from the anastomotic site is probably the oldest technique employed to occlude an artery to obtain hemostasis. This technique may cause major damage, especially to atheromatous arteries with, the risk of parietal plaque rupture, dissection and early thrombosis. Lately, cardiovascular surgeons have become concerned about these potential risks, particularly during manipulations of arterial conduits such as IMA. Different studies completed in this field showed that endothelial dysfunction due to extravascular clamping of the IMA was related to the type of clamp used and, more specifically, the type of jaw and the amount of pressure applied.[19-21] Histological assessments have documented a loss of endothelial coverage at the clamp site, resulting in endothelial dysfunction.[19] Structural lesions were also found to increase with the duration of IMA cross-clamping, endothelial damage being more severe after 15 minutes of device occlusion.[21] Bulldog jaws seemed to be the most traumatic cross-clamping devices whereas clamps covered with nylon fibers were more forgiving, allowing to some extent the preservation of functional islets of endothelial cells. The latter was shown to be sufficient to maintain normal vascular reactivity of the vessel and encourage endothelial cell regeneration.[20] However, the post-traumatic, regenerated endothelium does not always regain normal reactivity, and a post-regenerative dysfunctional state has been described.[22] This can favour post-operative spasm and the development of focal intimal hyperplasia.

Intracoronary Shunting

Intracoronary shunts (Fig. 1) have been used in coronary surgery since 1975 with satisfactory results.[9,23] This system has the dual theoretical advantage of keeping the anastomotic site bloodless but with effective perfusion of the distal segment of the coronary artery. This is sometimes necessary in beating heart off-pump surgery, particularly during occlusion of the right coronary artery or a dominant left coronary artery.[24] Furthermore, based on experimental and clinical studies, endovascular shunting can help to preserve left ventricular contractility during beating heart surgery and prevent further functional deterioration in patients with left ventricular dysfunction, or unstable angina.[25-27] Moreover, it can be beneficial in circumstances where longer procedural time is anticipated, such as a teaching session. Due to all these advantages, several authors have recommended the routine use of coronary shunts for beating heart surgery with excellent clinical results.[24,28,29] However, because they are in direct contact with the endothelium, shunts have to be approached with caution.

The effect of shunting on the endothelium was studied recently in healthy[30] and atheromatous[31] coronary arteries of swine. In healthy coronary arteries, comparison between 2 different shunts showed severe residual endothelial dysfunction regardless of the device type. Experiments performed on atherosclerotic arteries are of particular interest, since they more closely reproduce the clinical reality.[32] In a swine model, where coronary arteries were submitted

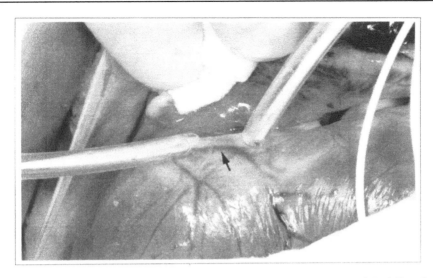

Figure 1. Coronary snares with threads wrapped around the coronary artery (experimental view). To occlude the vessel, the thread is put under tension and attached to the environment (e.g., retractors), or tightened on a tourniquet made usually with silicone tubing, allowing hemostasis of the anastomotic site (arrow).

percutaneously to balloon denudation 1 month prior to surgical experimentation with the shunts causing intimal hyperplasia resembling arteriosclerosis, the deployment of endovascular shunts did not induce any further injury. This was probably due to preexisting endothelial dysfunction in these arteries. The endoluminal trauma induced by the devices may not have any functional drawbacks, but their innocuousness still remains to be proven in further experimental studies focusing on the short- and long-term effects on endothelial reactivity and the occurrence of premature arteriosclerosis.

Suture Snaring

In 1979, Becker proposed snaring of the coronary arteries to obtain hemostasis.[10] More recently, different types of coronary snares (polypropylene 2/0 to 7/0, GoreTex, and silicone tape) to encircle the targeted artery with or without tourniquet have been popular among surgeons. Normally, the threads are kept under tension and fixed to the retractors or tightened on a tourniquet made of silicone tubing (Fig. 2). Some surgeons prefer to protect the coronary artery from the tubing by interposing a patch of pericardium or felt. Gundry et al in a case-matched study comparing a group of patients operated by CPB to a group operated on without CPB and without mechanical stabilization, found a 2-fold incidence of recatheterization in the beating heart group.[7] Among them, 20% required a second procedure, mainly percutaneous angioplasty. The majority of these interventions were made on previously-bypassed vessels. Only 7% of the CPB patients needed a second intervention, and the majority was performed on vessels non nongrafted previously.[7] These authors concluded that snares should not be applied for vessel stabilization because of the risk of causing proximal or distal anastomotic stenosis. Those cases were operated on before the coronary artery stabilizer era.

In a swine model, Perrault et al simulating hemostatic snaring (4/0 GoreTex thread on silicone tubing) of the coronary artery anastomotic site as used in clinical situations, showed that no endothelial dysfunction was created on normal coronary arteries.[33] However, in a recent study focusing on the ultrastructural aspect of coronaries arteries, Hangler et al described coronary artery lesions secondary to the application of hemostatic devices.[34] They performed

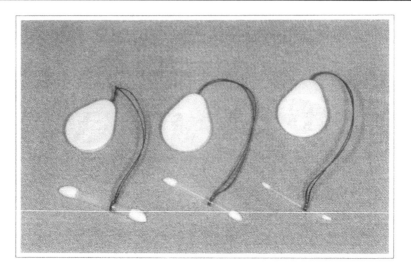

Figure 2. Intracoronary shunts of various sizes.

their experiments on the native heart of recipient patients in the process of being transplanted. After the initiation of CPB and under full anticoagulation, the heart was left beating, and the coronary arteries were occluded for 15 minutes using polypropylene buttressed with a piece of silicone tubing, an elastic silicone loop or the MyOcclude vessel occlusion device (Vascular Therapies, USSC, Elancourt, France). Scanning electron microscopy revealed that local occlusion of coronary arteries caused focal endothelial denudation, microthrombosis, and atherosclerotic plaque rupture. However, the lesions were less severe with a silicone loop or the MyOcclude devices. In conclusion, these authors advised against distal snaring to the arteriotomy.

Other complications have been described with snaring. Among them, distal atheromatous embolization, septal branch to heart chamber fistulas, and myocardial infarction.[35,36] In our experience, we have observed the development of early multifocal stenosis at 3 sites of coronary extravascular snaring 6 weeks after beating heart coronary artery bypass surgery in a diabetic patient. Since these patients are known to have extensive medial calcifications and a high risk of intimal hyperplasia development after arterial injury, a word of caution is mandatory during manipulation of these snares to minimize injury at the application site.[37]

Gas Jet Insufflation

Improved visualization of coronary artery anastomosis sites by of a gas jet (blower) insufflation was first reported in 1991 and was immediately controversial.[11,38] The risk of gaseous emboli seemed minimal with the use of CO_2,[39] but gas ejection under positive pressure was suspected to cause endothelial denudation. Indeed, this system was first described to remove the venous endothelium and abolish endothelium-dependent vasodilatation.[40] The deleterious effect on normal coronary arteries was further confirmed by histological study.[41] Loss of the endothelial layer may trigger coronary spasm and post-operative anastomotic thrombosis. Functional studies in organ chambers by Perrault et al confirmed these morphologic results, identifying a significant reduction of endothelium-dependent relaxation secondary to gas jet exposure.[12,42] Variations in flow, the duration of exposure and moisture of the gas may affect the degree of injury associated with this technique.

Extraluminal Balloon Occlusion

Another technique for coronary occlusion during beating heart coronary artery bypass surgery, using an extraluminal balloon occluder (Quest Medical, Inc., Allen, TX), was described recently.[12] It consisted of a blunt needle mounted on a silicone loop with a 3-cm long balloon of 3-mm diameter at its midsection. The occluder was applied through the myocardium under the coronary artery, and the balloon was inflated to obtain extravascular occlusion of the vessel. This device was studied in a swine model and compared to bulldog clamping and gas jet insufflation. Morphological (silver staining) and functional experiments (organ chambers) showed that cross-clamping and gas jet insufflation impaired endothelial function whereas application of the extravascular balloon occluder for 15 minutes provided effective hemostasis while preserving endothelial integrity. On histological assessment, morphological damage was minimal, and endothelial continuity was well-preserved. However, the balloon occluder occasionally caused superficial or intramyocardial hematoma due to its large diameter and the inflation-deflation cycle. For these reasons, the commercialization of this device was abandoned.

Clinical Studies

The goal of obtaining a bloodless operative field has been approached differently according to the surgical teams. In his preliminary report in 1985, Benetti[43] from Argentina operated on 30 selected patients with very good results. After general heparinization, he placed a double suture of 6/0 or 7/0 polypropylene proximally and distally to the anastomosis site. For coronary artery proximal and distal control, Pfister used small spring-loaded Bogart clamps or occasionally simple pressure with blunt forceps to control bleeding at the arteriotomy.[44] Buffolo,[45] a Brazilian surgeon and pioneer for revival of the beating heart technique, applied a 5/0 polypropylene suture around the coronary artery proximally and distally to the site selected for arteriotomy. The suture was then snared with a thin silicone tube, thereby allowing for a dry operative field. The results indicate that mortality was 2.5% and an 1% incidence of perioperative myocardial infarction. Calafiore tested a different technique of coronary occlusion during LAD artery grafting on the beating heart through a limited left anterior thoracotomy.[46] The LAD artery was occluded proximally and distally with a 4/0 polypropylene suture passed twice around the coronary vessel. To decrease potential arterial trauma, the suture was inserted through a short piece of silicone tubing, and only the proximal part of the vessel was gently snared. Subramanian[47] developed a similar technique but instead adopted silicone retractor tape to improve protection of the coronary artery (Quest Medical Inc.). Since 1995, he has also applied ischemic preconditioning (5 minutes of occlusion/5 minutes of reperfusion) before coronary blood flow interruption.[47] Like the preceding authors, Diegeler applied snares proximally and distally but with a soft vessel loop (Medipoint, Hamburg, Germany).[48] He has also used thread of 4/0 polypropylene over a piece of pericardium to protect the coronary vessel.[49] At the Montreal Heart Institute, most surgeons use silicone loop (Retract-O-Tape; Quest Medical Inc., Allen, TX) to isolate and occlude the target artery proximally and distally at the anastomotic site. Pedestals integrated on the coronary stabilizer for anchoring of the silicone loop allow better tension adjustment and may lessen the force applied to the vessel.[50] Others achieved a bloodless field with a single, proximal silicone vascular loop (Quest Medical, Inc.), leaving the distal artery undisturbed.[51] All these techniques were generally associated with good clinical short-term results.

Although snares are used by a majority of surgeons, some prefer local rinsing with warm saline solution and an atraumatic bulldog clamp placed proximally to the arteriotomy. Before completion of the anastomosis, the coronary artery is gently dilated with a 1 mm or 1.5-mm metal probe to treat possible residual coronary vessel spasm caused by the bulldog clamp.[52]

Other surgeons have combined different strategies to obtain hemostasis. Spooner[53] proposed a proximal silicone elastomere snare, alone, or combined with a distal silicone elastomere harness, or 4/0 polypropylene with Teflon buttresses combined with an intracoronary Flo-Restor (Bio-Vascular, St Paul, MN) or insertion of a Rivetti-Levinson intracoronary shunts (Integra Life Sciences, Plainsboro, NJ), if EKG changes developed. Visualization was improved with the Medtronic Clearview blower/mister (Medtronic, Inc., Minneapolis, Mn). In a series of 456 patients operated on with these techniques, he reported excellent short-term results with operative mortality and myocardial infarction rates of 0.32% and 0.8%, respectively.

Conclusions

Surgeons have the choice of several techniques and strategies for blood flow management during off-pump coronary artery grafting. Better knowledge of the effects of these devices on the biology of the arterial wall, and especially on endothelial reactivity, will provide better choices and guide in the development of an optimal technique.

Healthy coronary animal models with a normal endothelium, investigated typically in experimental studies, are more sensitive to surgical manipulations. The endothelium of atherosclerotic patients is generally dysfunctional, and occlusive devices may not significantly worsen the endothelial status of their coronary arteries. However, this latter fact does not exempt the surgeon from exploiting these devices with elementary caution, because of the risk of extensive endothelial denudation with subsequent spasm, dissection or embolism.

Moreover, an adequate anticoagulation regimen appears to be a major point for the optimization of clinical results. Off-pump patients should be considered at higher risk of thrombotic graft occlusion due to potential endothelial or arterial wall injury. Adequate intra-operative anticoagulation combined with a post-operative antiplatelet regimen is mandatory for every patient.[54,55]

Hemostatic systems will remain necessary in beating heart coronary artery bypass surgery as long as conventional suturing techniques will be used. Alternative techniques in the process of development such as sutureless, mechanical "one shot" anastomosis systems, are promising alternatives.

Ultimately, the optimal short- and long-term result of beating heart coronary artery surgery will depend not only on the hemostatic system chosen but also on the quality of cardiac stabilization.[56,57] The conjunction of the 2 techniques, in addition to better understanding of the consequences of coronary artery manipulation on vascular wall biology, will lead to the continued improvement of long-term results.

References

1. Carrel A. On the experimental surgery of the thoracic aorta and the heart. Ann Surg 1910; 52:83-95.
2. Sabiston DC. The coronary circulation. Johns Hopkins Med J 1974; 134:314-329.
3. Kolesov VI. Mammary artery-coronary artery anastomosis as method of treatment for angina pectoris. J Thorac Cardiovasc Surg 1967; 54:535-44.
4. Favaloro RG, Effler DB, Groves LK et al. Direct myocardial revascularization by saphenous vein graft. Present operative technique and indication. Ann Thorac Surg 1970; 10:97-111.
5. Benetti FJ, Naselli G, Wood M et al. Direct myocardial revascularization without extracorporeal circulation. Experience in 700 patients. Chest 1991; 100:312-16.
6. Pfister AJ. The safety of CABG without cardiopulmonary bypass. Ann Thorac Surg 1997; 64:590-91.
7. Gundry SR, Romano MA, Shattuck OH et al. Seven-year follow-up of coronary artery bypasses performed with and without cardiopulmonary bypass. J Thorac Cardiovasc Surg 1998; 115(6):1273-7.
8. Alessandrini F, Gaudino M, Glieca F et al. Lesions of the target vessel during minimally invasive myocardial revascularization. Ann Thorac Surg 1997; 64:1349-53.

9. Trapp VG, Bisarya R. Placement of coronary artery bypass graft without pump oxygenator. Ann Thorac Surg 1975; 19:1-9.

10. Becker RM. Local snares in coronary surgery. Lancet 1979; 1 (8109):213.

11. Teoh KHT, Panos AL, Harmantas AA et al. Optimal visualization of coronary artery anastomoses by gas jet. Ann Thorac Surg 1991; 52:564-5.

12. Perrault LP, Menasché P, Wassef M et al. Endothelial effects of hemostatic devices for continuous cardioplegia or minimally invasive operations. Ann Thorac Surg 1996; 62:1158-63.

13. Furchgott RF, Zawadzki JV. The obligatory role of endothelial cells in the relaxation of arterial smooth muscle by acetylcholine. Nature 1980; 288:373-6.

14. Furchgott RF. The Discovery of Endothelium-Dependant Relaxation. Circulation 1993; 87(suppl V):V-3-V-8.

15. Vanhoutte PM. The endothelium modulator of vascular smooth-muscle tone. N Engl J Med 1988; 319:512-3.

16. Vanhoutte PM. Endothelium and control of vascular function: State of the Art Lecture. Hypertension 1989; 13:658-67.

17. Vanhoutte PM, Rimele TJ. Role of the endothelium in the control of vascular smooth muscle function. J Physiol 1983; 78:681-6.

18. He GW, Yang CQ. Comparaison among arterial grafts and coronary artery. An attempt at functional classification. J Thorac Cardiovasc Surg 1995; 109:707-15.

19. Fonger JD, Yang XM, Cohen RA et al. Impaired relaxation of the human mammary artery after temporary clamping. J Thorac Cardiovasc Surg 1992; 104:966-71.

20. Fonger JD, Yang XM, Cohen RA et al. Human mammary artery endothelial sparing with fibrous jaw clamping. Ann Thorac Surg 1995; 60:551-5.

21. Kuo J, Ramstead K, Salih V et al. Effect of vascular clamp on endothelial integrity of the internal mammary artery. Ann Thorac Surg 1993; 55:923-6.

22. Shimokawa H, Aarhus LL, Vanhoutte PM. Porcine coronary arteries with regenerated endothelium have a reduced endothelium-dependant responsiveness to aggregating platelets and serotonin. Circ Res 1987; 61:256-70.

23. Franzone AJ, Wallsh ES, Stertzer H et al. Reduced incidence of intraoperative myocardial infarction during coronary bypass surgery with use of intracoronary shunt technique. Am J Cardiol 1977; 39(7):1017-20.

24. Levinson MM, Fooks GS. Coronary grafting using a temporary intraluminal shunt instead of heart-lung bypass. Ann Thorac Surg 1995; 60:1800-1.

25. Dapunt OE, Raji MR, Jeschkeit S et al. Intracoronary shunt insertion prevents myocardial stunning in a juvenile porcine MIDCAB model absent of coronary artery disease. Eur J Cardiothorac Surg 1999; 15:173-9.

26. Lucchetti V, Capasso F, Caputo M et al. Intracoronary shunt prevents left ventricular function impairment during beating heart coronary revascularization. Eur J Cardiothorac Surg 1999; 15:255-9.

27. Ricci M, Karamanoukian HL, D'Ancona G et al. Survey of resident training in beating heart operations. Ann Thorac Surg 2000; 70 (2):479-82.

28. Rivetti LA, Gandra SMA. Initial experience using an intraluminal shunt during revascularization of the beating heart. Ann Thorac Surg 1997; 63:1742-47.

29. Arai H, Yochida T, Izumi H et al. External shunt for off-pump coronary artery bypass grafting: Distal coronary perfusion catheter. Ann Thorac Surg 2000; 70:681-2.

30. Chavanon O, Perrault LP, Menasché P et al. Endothelial effects of hemostatic devices for continuous cardioplegia or minimally invasive operations. Updated in 1999. Ann Thorac Surg 1999; 68:1118-20.

31. Perrault LP, Desjardins N, Nickner C et al. Effects of occlusion devices for minimally invasive coronary artery bypass surgery on coronary endothelial function of atherosclerotic arteries. Heart Surg Forum 2000; 3:287-92.

32. Morelos M, Amyot R, Picano E et al. Effect of coronary bypass and cardiac valve surgery on systemic endothelial function. Am J Cardiol 2001; 87:364-6.

33. Perrault LP, Menasché P, Bidouard JP et al. Snaring of the target vessels in less invasive bypass operations does not cause endothelial dysfunction. Ann Thorac Surg 1997; 63:751-5.

34. Hangler HB, Pfaller K, Antretter H et al. Coronary endothelial injury after local occlusion on the human beating heart. Ann Thorac Surg 2001; 71:122-7.
35. Izzat MB, Yim APC, El-Zufari H. Snaring of a coronary artery causing distal atheroma embolization. Ann Thorac Surg 1998; 66:1806-8.
36. Tanemoto K, Kuroki K, Kanaoka Y et al. Septal branch right ventricular fistula: A complication in coronary artery snaring. Ann Thorac Surg 1999; 68:246-8.
37. Demaria RG, Fortier S, Carrier M et al. Early multifocal stenosis after coronary artery snaring during OPCAB in a diabetic patient. J Thorac Cardiovasc Surg 2001; 122(5):1044-5.
38. Poulton TJ. Visualisation of coronary artery anastomoses by gas jet [letter]. Ann Thorac Surg 1992; 54:598-9.
39. Sasaguri S, Hosoda Y, Yamamoto S. Carbon dioxid gas blow for the safe visualization of coronary artery anastomosis [letter]. Ann Thorac Surg 1995; 60:1861.
40. Bjorling DE, Saban R, Tengowski MW et al. Removal of venous endothelium with air. J Pharmacol Toxicol Methods 1992; 28:149-57.
41. Burfeind Jr WR, Duhaylongsod FG, Annex BH et al. Height-flow gas insufflation to facilitate MIDCABG: Effects on coronary endothelium. Ann Thorac Surg 1998; 66:1246-9.
42. Perrault LP, Menasché P, Vanhoutte PM. High-flow gas insufflation to facilitate MIDCAB [letter]. Ann Thorac Surg 1999; 67:893.
43. Benetti FJ. Direct coronary surgery with saphenous vein bypass without either cardiopulmonary bypass or cardiac arrest. J Cardiovasc Surg 1985; 26:217-22.
44. Pfister AJ, Zaki MS, Garcia JM et al. Coronary artery bypass without cardiopulmonary bypass. Ann Thorac Surg 1992; 54:1085-92.
45. Buffolo E, Silva de Andrade JC, Rodrigues Branco JN et al. Coronary artery bypass grafting without cardiopulmonary bypass. Ann Thorac Surg 1996; 61:63-6.
46. Calafiore AM, Giammarco GD, Teodori G et al. Left anterior descending coronary artery grafting via left anterior small thoracotomy without cardiopulmonary bypass. Ann Thorac Surg 1996; 61(6):1658-63.
47. Subramanian VA, McCabe JC, Geller CM. Minimally invasive direct coronary artery bypass grafting: Two-year clinical experience. Ann Thorac Surg 1997; 64(6):1648-53.
48. Diegeler A, Falk V, Walther T et al. Minimally invasive coronary artery bypass surgery without extracorporeal circulation. N Engl J Med 1997; 336:1454.
49. Diegeler A, Falk V, Matin M et al. Minimally invasive coronary artery bypass grafting without cardiopulmonary bypass: Early experience and follow-up. Ann Thorac Surg 1998; 66(3):1022-5.
50. Cartier R. Systematic off-pump coronary artery revascularization: Experience of 275 cases. Ann Thorac Surg 1999; 68(4):1494-7.
51. Turner Jr WF. "Off-pump" coronary artery bypass grafting: The first one hundred cases of the Rose city experience. Ann Thorac Surg 1999; 68:1482-85.
52. Tasdemir O, Vural KM, Karagoz H et al. Coronary artery bypass grafting on the beating heart without the use of extracorporeal circulation: Review of 2052 cases. J Thorac Cardiovasc Surg 1998; 116(1):68-73.
53. Spooner TH, Hart JC, Pym J. A two-year, three-institution experience with the Medtronic Octopus: Systematic off-pump surgery. Ann Thorac Surg 1999; 68(4):1478-81.
54. Mariani MA, Gu YJ, Boonstra PW et al. Procoagulant activity after off-pump coronary operation: Is the current anticoagulation adequate? Ann Thorac Surg 1999; 67:1370-75.
55. Hangler HB, Pfaller K, Antretter H et al. Coronary endothelial injury after local occlusion on the human beating heart. Ann Thorac Surg 2001; 71:127.
56. Poirier NC, Carrier M, Lesperance J et al. Quantitative angiographic assessment of coronary anastomoses performed without cardiopulmonary bypass. J Thorac Cardiovasc Surg 1999; 117(2):292-7.
57. Perrault LP, Nickner C, Desjardins N et al. Effects on coronary endothelial function of the Cohn stabilizer for beating heart bypass operations. Ann Thorac Surg 2000; 70(3):1111-4.

CHAPTER 8

Systematic OPCAB Surgery for Multivessel Disease with the CoroNéo Cor-Vasc Device

Raymond Cartier

Introduced in the mid-1960s for single vessel, OPCAB surgery is currently applicable to multivessel disease.[1-4] The advent of mechanical stabilizers has been, without any doubt, a giant leap in the evolution of this technique. Needless to say that 40 years of collective experience with the coronary artery grafting technique have contributed to settling the modern foundation of OPCAB surgery, allowing its use in a majority of patients.[5] Moreover, aging of the surgical population and the more aggressive use of coronary angioplasty have increased the surgical challenge. Surgeons are tempted to reconsider less invasive surgical approaches to cope with these new rules. Over the last few years, several randomized trials have been performed and more than 12,000 cases from the Society of Thoracic Surgery database have been reported and published. Avoidance of cardiopulmonary bypass (CPB) is beneficial for several reasons but is indubitably more demanding technically.

This is an overview of our experience with the systematic use of OPCAB surgery along with a recent literature review on the topic.

Patient Population

Between September 1996 and April 2002, 750 OPCAB surgeries were undertaken at the Montreal Heart Institute. This was a single surgeon's experience and represented 95% of all coronary artery bypass graftings (CABG) cases done during that period (98% in the last 4 years). All cases were performed with the same surgical technique (see below). The main contraindications were preoperative unstable hemodynamics, a very deep intramyocardial left anterior descending (LAD) artery, and reoperations where dense posterior adhesion could not allow safe dissection. These cases were compared to a series of 1,444 patients operated on with CPB between January 1997 and December 1998 at the Montreal Heart Institute. Six surgeons were involved at that time (including the current author). Emergency procedures (life-saving procedures and patients operated on less than 12 hours after the diagnosis for unstable conditions) performed were excluded (82 cases representing 5.6% of the entire cohort) since these cases were almost always subjected to CPB.

The OPCAB Surgical Technique

The surgical technique has been described in Chapter 6. Briefly, a compression type device (Cor-Vasc retractor-stabilizer, CoroNéo, Montreal, QC, Canada) was employed to achieve coronary artery stabilization.[5] The 8 degrees of freedom arm of the apparatus adaptable to 4 different stabilizers allowed access to all coronary territories (Fig. 1). Two of the stabilizers were

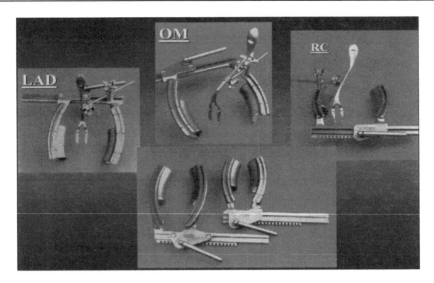

Figure 1. Cor-Vasc device with the different coronary artery stabilizers for the specific targeted vessels.

"push type", pressurecontact devices, one at 100° and the other at 160° acute angle. The other two were "pull type" (220°) with leftward or rightward orientation, mainly designed to access the posterior territory. Most (99%) of the procedures were approached through a median sternotomy. The distal right coronary artery (RCA) was accessed by rotating the table away from the surgeon (20° to 25°) and by implanting 2 pericardial traction sutures, one above the inferior vena cava (IVC) and the other in front of the right atrium. These tractions transposed the RCA on the mid-line, facilitating access. The anterior territory was exposed with the table slightly in the "reversed" Trendelenburg position, rotated towards the surgeon (15° to 20°). Generally, a single pericardial traction suture was anchored 1 centimeter above the phrenic nerve at the area of maximum pericardial deformation by the apex of the heart. This brought the LAD artery on the mid-line, and also gave access to the diagonal arteries. The proximal posterior descending artery was reached similarly to the LAD artery but with generous Trendelenburg positioning.

To access the circumflex territory, the table was kept in the Trendelenburg position and rotated towards the surgeon (30°). Upward traction was applied to the free edge of the pericardium, and the bottom of the pericardial sac was kept away from the heart using the aspirator tip. The apex of the heart was pulled gently and 4 deep pericardial traction sutures were anchored between the left superior pulmonary vein and the IVC, close to the pericardial reflection. This maneuver allowed verticalization of the apex of the heart without any further manipulation. Then, angulated stabilizers were used to isolate the obtuse marginal arteries.

Coronary blood flow interruption was achieved by gentle snaring (silicone wires) close to the arteriotomy. Loops were fixed on the stabilizer itself with minimal tension. Distally, only the posterior wall of the vessel was circled to minimize potential arterial trauma.

No pleural heart herniation or right heart bypass was used. In cases of very dilated hearts, a vertical pericardial incision was made at the acute angle of the right heart down to the IVC. This lowered the base of the heart in the chest cavity, increasing the surgical working space on the posterior territory.

Surgical Strategies

Surgical strategies have already been discussed in Chapter 6 and will be mentioned here briefly. To maintain the conversion rate to CPB as low as possible, we observed, early in our experience, a very strict revascularization policy consisting of always first bypassing the vessel with the culprit lesion. This vessel, generally due to its well-developed collateral network, tolerated temporary occlusion very well. Proximal anastomosis, in case of a saphenous or free arterial conduit, was then carried out, reestablishing forward blood flow into the graft and collateral vessels. This provided circulatory backup during subsequent revascularization of less critically stenotic arteries. Lately, we have used skeletonized internal thoracic arteries extensively, which in our experience facilitated sequential bypasses because of the extra length provided.

Anesthetic Management

Anesthetic management during OPCAB surgery has been described in Chapter 6.[2] Patients were initially given intravenous fluid to maintain central venous pressure around 10 mm Hg prior to surgical manipulation. Systemic arterial pressure was maintained with α-agonist drugs, either by bolus or perfusion, to target systolic pressure above or equal to 100 mmHg during maneuvers. Beta-blockers were administered only when heartbeat was over 85 per minute, to avoid increased myocardial oxygen consumption and ease coronary grafting. Intravenous nitroglycerine was normally given whenever signs of myocardial ischemia or pulmonary pressure elevation were present. The intra-aortic balloon pump was applied occasionally prior to surgery to stabilize the patient's hemodynamics when indicated. Hypothermia during the procedure was avoided with an intravenous heating device and heating mattress.

Inflow Maneuver

De novo pulmonary hypertension developing during the procedure (generally due to temporary left ventricular ischemia) that was not responding to intravenous nitroglycerine was managed by intermittent partial or total cross-clamping of the IVC. This was generally achieved by progressing snaring of the IVC,[6,7] which instantly decreased the ventricular preload and helped reestablishing left ventricular dynamics. Inflow interruption of the IVC was never maintained for more than a few minutes. This maneuver helped to avoid emergency conversion.

The CPB Surgical Technique

In the CPB cohort, the extracorporeal circulation was established with membrane oxygenator, roller pump, and moderate hypothermia (32°C). Cardiac arrest was achieved with either cold (< 20°C) or tepid (25°C) hyperkalemic (24 mEq KCl) antegrade cardioplegia.

Statistics

Computerized statistical analysis was performed with SPSS software (SPSS Inc., Chicago, IL, USA) for Windows. The clinical data are expressed as mean and standard deviation or median and interquartile range. For categorical variables, the chi-square test was used, whereas discrete variables were evaluated by one way analysis of variance. Statistical significance was established at $p < 0.05$. Univariate and step-wise logistic regression analysis were performed using the same software to determine risk factors for categorical variables, such as operative mortality and post-operative events.

Results

The demographics and preoperative characteristics are reported in Table 1. Both cohorts were comparable, although a history of cardiac failure was more frequent in the CPB group, and the preoperative use of an intra-aortic balloon pump (IABP) more common in OPCAB patients. Intra-operative technical data were similar for both groups, although, on average, more grafts/patient and more grafts/territory were performed in OPCAB patients. Total ischemic time (cumulative local ischemic period in cases of OPCAB patients and aortic cross-clamp time in CPB patients) was lower in the off-pump cohort (Table 2). The conversion rate in OPCAB patients remained inferior to 1% during the entire period.

Post-operative outcome is reported in Table 3. The rates of peri-operative myocardial infarction (MI), stroke, atrial fibrillation, reoperation for bleeding and hospital stay were comparable for both groups. However, the post-operative use of IABP, maximal CK-MB count, transfusion rate, amount of blood products given per transfused patient, and hospital stay beyond 7 days were significantly higher in the conventional group. Interestingly, even though OPCAB patients started with a lower hemoglobin (Hb) count and had fewer transfusions, they left the hospital with a higher Hb count (Table 4). Post-operative creatinine was also higher in the conventionally-operated patients (Table 4).

Table 1. Demographics and preoperative risk factors for off-pump (OPCAB) and conventional surgery (CPB)

	OPCAB	CPB	p Value
Age	64 ± 10	63 ± 9	NS
Sex ratio	3.54	3.71	NS
Diabetes	25%	21%	NS
HBP	47%	54%	NS
Cigarette smoking	32%	29%	NS
COPD	11%	9%	NS
Redo surgery	7%	7%	NS
Previous MI			
< 30 days	16%	18%	NS
> 30 days	44%	42%	NS
Previous CVA	9%	7.3%	NS
Carotid bruit	16%	13%	NS
PVD	19%	14%	NS
Cardiac insufficiency	9%	15%	< 0.05
Unstable angina	70%	72%	NS
LVEF	54% ± 13%	54% ± 15%	NS
LVEF < 40%	14%	17%	NS
Left main stem > 50%	30%	28%	NS
Pre-operative IABP	7%	3.5%	< 0.05
Territory involved	2.7 ± 0.6	2.6 ± 0.54	NS

HBP= High blood pressure; COPD= Chronic obstructive pulmonary disease; MI= Myocardial infarction; CVA= Cerebrovascular accident; PVD= Peripheral vascular disease; LVEF= Left ventricular ejection fraction; IABP= Intra-aortic balloon pump

Table 2. Technical data for off-pump (OPCAB) and conventional surgery (CPB)

	OPCAB	CPB	p Value
Grafts/patient	3.16 ± 0.94	2.91 ± 0.67	0.006
Ischemic time (min)	30 ± 11	43 ± 16	0.001
Grafts/territory	1.18 ± 0.26	1.14 ± 0.34	0.005
Complete revascularization	93%	NA	
Conversion rate	0.4%		

Table 3. Peri-operative events for off-pump (OPCAB) and conventional surgery (CPB)

	OPCAB	CPB	p Value
Peri-operative MI	3.4%	4.2%	NS
Operative mortality	1.4%	1.9%	NS
Hemorrhage	4.3%	4.7%	NS
Atrial fibrillation	27%	30%	NS
CVA (TIA)	0.8%	1.0%	NS
Post-operative IABP	0.7%	2.9%	0.01
Inotropic support	36%	32%	NS
Hospital stay (days)	6.8 ± 6.1	6.7 ± 6.23	NS
CK-MB (IU/L) Max	19 ± 43	42 ± 51	0.001
Creatinine (50 mmol/L)	11%	10.5%	NS
Hospital stay > 7 days	22%	30%	0.001
Transfusion	27%	62%	0.001
BP/transfused patient	4.2 ± 6.1	6.8 ± 8.6	0.001

MI= myocardial infarction "Q" and "non Q"; CVA= cerebrovascular accident; TIA = transient ischemic attack; IABP= intra-aortic balloon pump; CK-MB= Creatine kinase (myocardium specific); BP= Blood products

Table 4. Biochemistry for off-pump (OPCAB) and conventional surgery (CPB)

	OPCAB	CPB	p Value
Hemoglobin (g/L)			
Pre-operative	131 ± 21	136 ± 16	< 0.05
Post-operative	107 ± 14	102 ± 18	< 0.05
Creatinine (mmol/L)			
Pre-operative	101 ± 58	105 ± 51	ns
Post-operative	102 ± 53	123 ± 73	< 0.05

Risk Factor Analysis for Operative Mortality

Preoperative independent risk factors that were predictive of operative mortality are displayed in Table 5. Univariate analysis revealed that age, sex, a history of cardiac insufficiency, previous stroke, the use of an IABP prior to surgery, left main stem stenosis, redo operation, chronic renal insufficiency, and decreased left ventricular function were significant risk factors.

Stepwise logistic regression analysis upheld age, female sex, renal insufficiency, redo surgery, left main stenosis, and left ventricular function as independent predictive risk factors. When the type of surgery was furthered included in the logistic regression model it did not reach statistical significance (p=0.22). Separate analysis performed in the CPB cohort disclosed that bypass time (p=0.04, OR=1.012) but not aortic cross-clamping(XC) time (p=0.5) was an independent risk factor for post-operative mortality. In the OPCAB cohort, total ischemic time did not correlate with operative mortality (p=0.75). Similarly, linear regression analysis revealed that in the OPCAB cohort, cumulative ischemic time, which averaged 31±12 minutes (7-89 min), did not correlate with maximum CK-MB count. In CPB patients XC time did not correlate with the rise in post-operative CK-MB count.

We also performed risk factor analysis using as a dependent variable a cumulative complication index of operative mortality, peri-operative MI, post-operative IABP, and hospital stay >7 days. These results appear in Table 6. CPB was identified as a contributing risk factor for the entire cohort and also in patients with left main disease, and left ventricular dysfunction (left ventricular ejection fraction < 40%). In the elderly (> 70 years old), a trend was still present but was not significant, due to the smaller patient population numbers were small. Outcome was not influenced by the surgical technique in diabetic patients and women.

Mid-Term Follow-Up

We recently carried out a survey of the last 340 consecutive patients who had a minimum of 12 months follow-up after surgery (average follow-up 24 ±9.5 months). Among them 9.5% were readmitted for cardiac-related problems. Of these 1.5% were readmitted for new onset of MI, 0.6% for unstable angina, and 2.4% for cardiac failure. New onset of cerebrovascular

Table 5. Preoperative predictive factors for operative mortality in both groups using univariate (UA) and binary logistic regression (LRA) analysis with odds ratios (OR)

	UA	LRA	(OR)
Age	0.005	0.001	(1.08)
Sex*	0.001	< 0.001	(3.75)
Diabetes	0.47		
HBP	0.945		
COPD	0.68		
IABP (preop)	0.001	0.56	(1.37)
Heart failure	0.014	0.128	(1.9)
MI (< 30 days)	0.55		
PVD	0.43		
Stroke	0.025	0.015	(3.83)
LVEF	0.014	0.001	(0.19)
Unstable angina	0.328		
Left main	0.005	0.005	(2.69)
IABP	0.158		
Redo surgery	0.002	0.001	(4.35)
Renal insufficiency	0.001	< 0.001	(6.41)
Surgical technique	0.184	0.22	(1.6)

*Reference category = male; HBP= High blood pressure; COPD= Chronic Obstructive pulmonary disease; ABP= Intra-aortic balloon pump; MI= Myocardial infarction; PVD= Peripheral vascular disease; LVEF= Left ventricular ejection fraction

Table 6. Risk-adjusted analysis for a cumulative complication index including hospital mortality (30 days), myocardial infarction (CK-MB > 100), postoperative use of IABP, and hospital stay > 7 days using stepwise binary regression.

All Patients	p Value	O.R.
Age	< 0.001	1.04
Sex	< 0.001	1.61
LVEF	0.015	1.37
IABP (pre-operative)	< 0.001	9.1
PVD	< 0.001	2.1
Surgical technique*	**0.007**	**1.4**
> 70 Years		
IABP (pre-operative)	< 0.001	6.4
PVD	0.015	2.1
Surgical technique	**0.068**	**1.4**
Left Main		
Age	0.001	1.05
IABP (pre-operative)	0.001	3.1
PVD	< 0.001	3.1
Surgical technique	**0.025**	**1.6**
Left Ventricular Ejection Fraction < 40%		
Age	0.003	1.05
Sex	< 0.001	0.65
IABP (pre-operative)	0.001	8.57
PVD	< 0.001	2.39
Surgical technique	**0.022**	**1.4**
Women		
IABP (pre-operative)	< 0.001	14.1
PVD	0.091	1.6
Diabetics		
IABP (pre-operative)	< 0.001	5.4
LVEF	0.009	0.13
SEX	0.09	1.52
PVD	0.047	1.7
Age	0.101	1.02

*Reference category: OPCAB=0. LVEF= Left ventricular ejection fraction; IAPB= Intra-aortic balloon pump; PVD= Peripheral vascular disease

accident occurred in 0.6% of the cohort. A coronary angiogram was performed in 15 of them (4.4%) for recurrent angina, and 8 (2.4%) underwent a percutaneous coronary angioplasty. Only 1 patient (0.3%) needed a coronary artery reoperation for the progression of native disease. Graft patency is reported in Table 7. By and large, patency was 84% (74% with no

Table 7. Graft patency according to the Fitzgibbon classification

	A	B	0	Patency*
LAD	11	2	2	87% (73%)
Diagonal	5	0	2	71% (71%)
OM	5	1	3	67% (56%)
PDA	3	0	0	100% (100%)
RC	4	1	0	100% (80%)
PL	3	0	0	100% (100%)
TOTAL	31	4	7	
	(74%)	(10%)	(16%)	

*Percentages in parentheses represent type A patency. LAD = Left anterior descending artery; OM = Optuse marginal; PDA = Posterior descending artery; RC = Right coronary artery; PL = Posterolateral coronary artery

significant stenosis). These results are quite encouraging considering that only symptomatic patients were recatheterized. For comparison in the Fitzgibbon[8] study, where all patients were studied systematically regardless of their symptoms, 88% of the grafts were patent, and 82% were rated grade A (no significant stenosis).

Eleven patients (3.2%) died during follow-up, 3 of them from cardiac causes. Acturarial survival was 96% at 3 years.

Comments

Although several authors have reported doing most of their coronary cases off-pump, there are very few reports on the systematic approach for OPCAB surgery. Most of the prospective, randomized studies have been performed on highly-selected patients, and generally do not represent the daily practice of cardiac surgeons. For instance, these trials excluded redo surgery, patients with recent MI or poor ejection fraction. In 2001, Van Dijk and colleagues reported on 281 patients randomly assigned to off- or on-pump surgery.[9] These patients were randomized in 3 hospitals in the Netherlands over a period of 17 months. They excluded cases of emergency or concomitant major surgery, Q-wave MI in the last 6 weeks, poor ejection fraction and patients unlikely to complete 1 year of follow-up. Angelini and the Bristol group in 2002 published the combined report of the BHACAS 1 and 2 trials (Beating Heart Against Cardioplegic Arrest Studies).[10] In the first trial, 200 out of 538 consecutively admitted patients (37%) were randomized, whereas in the second trial, 201 of 320 consecutive patients (63%) were assigned. Diegler, in a randomized trial comparing stenting and MIDAB for single LAD artery disease randomized 220 consecutive patients.[11] This was a great achievement but not representative of regular surgical practice. The report of Nathoe and colleagues from the Octopus Study Group on off-pump surgery in low-risk patients did not mention the rate of enrollment.[12]

In general, these trials did not find any improvement in favour of OPCAB surgery in regard to operative mortality due to the low-risk nature of the cohort studied. The peri-operative infarction rate was similar with both techniques, meaning that patient anatomy rather than the surgical approach was the most important factor. The similar results achieved in operative mortality, peri-operative MI and stroke confirm the safety and security of off-pump surgery. Nevertheless, reperfusion injury (as measured by lower CK-MB and troponin counts) was significantly attenuated in OPCAB surgery along with decreased transfusions, intensive care unit stay, and intubation time. All these studies had on thing in common: the cohorts examined were relatively small. Cleveland and colleagues examined risk-adjusted mortality and morbidity in off and on–pump surgery in a cohort of 118,140 CABG procedures, including 11,717 off-pump cases, distributed in 126 experienced centers.[13] Risk-adjusted operative mortality was significantly lower in the off-pump group (2.3% vs 2.9%) as was the occurrence of major complications (10.62% vs 14.15%) such as stroke, coma, mechanical ventilation >24 hours, renal failure (defined as a 50% increase over the preoperative creatinin baseline level or dialysis), mediastinitis, and reoperation for bleeding. Statistical significance was achieved due to the large numbers involved; but, still, the cases were selected.

Our cohort of systematic OPCAB patients included all cases done during the last 6 years, regardless of their anatomy, surgical indication, or age range. Only patients who presented very unstable hemodynamics necessitating genuine inotrope use during anesthetic induction were generally not considered for the technique, although 3 patients in cardiogenic shock were operated on successfully off-pump during that period. This cohort reflects the daily activity of a cardiac surgeon in the current era. Hard end-points, such as operative mortality, peri-operative stroke, and MI were comparable to conventional surgery, as shown in the randomized study. Benefits were observed in terms of myocardial preservation, blood product transfusions, prolonged hospital stay, and post-operative IABP. When we risk-adjusted operative mortality for the independent risk factors identified through logistic regression analysis, no definite trend was noted in favour of the off-pump technique. However, when the cumulative complication index was taken as an endpoint, significant findings were made. The cumulative complication index that we chose referred to simple and relatively hard endpoints.

Operative mortality and peri-operative MI are both strong markers of the quality of the surgical act itself, whereas the post-operative use of IABP and prolonged hospital stay beyond 7 days are good markers for cardiac or other complications that significantly impact on post-operative outcome. The off-pump technique was found to be globally beneficial for the entire cohort, but more specifically for patients with left main disease and left ventricular dysfunction. In these patients with a precarious anatomy, the myocardium is more susceptible to reperfusion injury during conventional surgery, and avoiding global ischemia could be helpful. Other groups have reported similar results.[14-16] Avoiding CPB was also found to be beneficial in the elderly (> 70 years). Aging has been identified as a poor prognosis factor in conventional cardiac surgery, with these patients being more susceptible to the side-effects of CPB.[17] Conversely, women and diabetic patients do not have any specific advantage with the technique.

Female gender has been recognized as an independent risk factor in coronary artery surgery. Woods and colleagues,[18] in a study gathering 3,582 men and 1,742 women, found that in general women had more comorbidities and higher nonadjusted operative mortality than men. Regression analysis revealed that sex itself was not an independent risk factor. However, when occurrence of intra-operative events, low cardiac output syndrome and length of stay were considered, female gender was identified as an independent risk factor. These findings tend to corroborate a common belief that the poorer prognosis of women is related to anatomical problems such as a smaller coronary network, which decreases the technical success of the procedure. Brown et al examined the outcome of off-pump coronary surgery in women.[19] Using the Healthcare Company Casemix Database, they compiled clinical data on cardiovascular patients who underwent coronary surgery from January 1998 to June 2001 in 78 different hospitals. They compared 14,240 women who underwent on-pump surgery to 2,631 women who had off-pump surgery, for operative mortality and 13 procedure complications. They took into consideration 35 variables that covered preoperative characteristics and comorbid conditions. The major findings were lower operative mortality in the off-pump cohort (3.12 vs 3.90; p=0.052) along with reduced length of hospital stay and respiratory complications. However, occurrence of the 12 other complications was not influenced by the choice of the surgical technique. In our cohort, the type of surgery and patient gender impacted the incidence of

Table 8. Operative mortality according to the technique used and the patient gender

	Men (n)	Women (n)	p
Total	1.5% (1755)	4.9% (439)	< 0.001
Off-pump	1.2% (590)	2.9% (160)	0.252
On-pump	1.7% (1140)	5.8% (312)	< 0.001

operative mortality (Table 8). By and large, operative mortality was lower in men, although the difference was significant only in the on-pump cohort. Nevertheless, as reported by Brown, when female gender alone was considered, a positive trend in favour of off-pump surgery was observed with the off–pump approach (2.9% vs 5.8%, p=0.16). Statistical significance was not reached due to the small numbers involved.

Diabetes has been identified as an independent risk factor in coronary artery surgery.[20,21] Peri-operative control of glycemia seems to be a key element and perhaps more important than the surgical technique itself since diabetes has been established as a predictive risk factor for a poorer outcome rather than a predictor for operative mortality.[22] In our cohort a similar outcome was observed regardless of the presence of diabetes, which suggests that the surgical technique itself was not a relevant factor in morbidity related to the procedure. Left main stem stenosis (LMS) has long been recognized as a risk factor for coronary artery surgery. According to the STS the presence of LMS still increases the operative mortality by 1.5-fold (4.7% vs 2%”).[23] Several reports have confirmed the security of off-pump surgery in patients with LMS.[14-16] Although these authors have recorded a decreased operative mortality in patients operated off-pump, the difference did not reach significance. In the current cohort, operative mortality favoured off-pump (2.2% vs 3.4%, p=ns) as did as the cumulative complication index. Our data tend to confirm that the off-pump technique can be safely used in the presence of left main disease.

Conclusion

So far, our current experience with the systematic OPCAB approach has been favourable. It was shown to be safe and efficient. Although we do not state that this approach should be adopted universally it could be used successfully. Long-term data are needed to confirm these encouraging results.

References

1. Kolesof VI. Mammary-coronary artery anastomosis as method of treatment for angina pectoris. J Thorac Cardiovasc Surg 1967; 54:535-44.
2. Cartier R, Blain R. Revascularization of the circumflex artery without cardiopulmonary bypass: Technical aspect and short-term results. Ann Thorac Surg 1999; 68:94-9.
3. Buffolo EA, Andrade JCS, Branco JNR et al. Myocardial revascularization without extracorporeal circulation: Seven-year experience in 593 cases. Eur J Cardiothorac Surg 1990; 4:504-8.
4. Benetti FJ, Naselli G, Wood M et al. Direct myocardial revascularization without extracorporeal circulation: Experience of 700 patients. Chest 1991; 100:310-6.
5. Cartier R, Brann S, Dagenais F et al. Systematic off-pump coronary artery revascularization in multivessel disease: Experience of 300 cases. J Thorac Cardiovasc Surg 2000; 119:221-9.
6. Dagenais F, Cartier R. Pulmonary hypertension during beating heart coronary surgery: Intermittent inferior vena cava snaring. Ann Thorac Surg 1999; 68:1094-5.
7. Couture P, Denault AY, Limoges P et al. Mechanisms of hemodynamic changes during off-pump coronary artery bypass surgery. Can J Anaesth 2002; 49(8):835-49.

8. Fitzgibbon GM, Kafka HP, Leach AJ et al. Coronary bypass graft fate and patient outcome: Angiographic follow-up of 5065 grafts related to survival and reoperation in 1,388 patients during 25 years. J Am Coll Cardiol 1996; 28:616-26.
9. van Dijk D, Nierich AP, Jansen EW et al. for the Octopus Study Group. Early outcome after off-pump versus on-pump coronary bypass surgery: Results from a randomized study. Circulation 2001; 104:1761-6.
10. Angelini GD, Taylor FC, Reeves BC et al. Early and midterm outcome after off-pump and on-pump surgery in Beating Heart Against Cardioplegic Arrest Studies (BHACAS 1 and 2): A pooled analysis of two randomized controlled trials. Lancet 2002; 359:1194-9.
11. Diegler A, Thiele H, Falk V et al. Comparison of stenting with minimally invasive bypass surgery for restenosis of the left anterior descending coronary artery. N Engl J Med 2002; 347(8):561-6.
12. Nathoe HM, van Dijk D, Jansen EW et al. from the Octopus Study Group. A comparison of on-pump and off-pump coronary bypass surgery in low-risk patients. N Engl J Med 2003; 348(5):394-402.
13. Cleveland JC, Shroyer ALW, Chen AY et al. Off-pump coronary artery bypass grafting decreases risk-adjusted mortality and morbidity. Ann Thorac Surg 2001; 72:1282-9.
14. Yeatman M, Caputo M, Ascione R et al. Off-pump coronary artery bypass surgery for critical left main stem disease: Safety, efficacy and outcome. Eur J Cardiothorac Surg 2001; 9(3):239-44.
15. Meharwal ZS, Trehan N. Is off-pump coronary artery bypass surgery safe for left main coronary artery stenosis? Indian Heart J 2001; (3):314-8.
16. Brann S, Martineau R, Cartier R. Left main coronary artery stenosis: Early experience with surgical revascularization without cardiopulmonary bypass. J Cardiovasc Surg (Torino) 2000; 41(2):175-9.
17. Kilo J, Czerny M, Zimpfer D et al. Predictors of perioperative mortality after coronary artery bypass grafting in the elderly. Thorac Cardiovasc Surg 2003; 51(1):33-7.
18. Woods SE, Noble G, Smith JM et al. The influence of gender in patients undergoing coronary artery bypass graft surgery: An eight-year prospective hospitalized cohort study. J Am Coll Surg 2003; 196(3):428-34.
19. Brown PP, Mack MJ, Simon AW et al. Outcomes experience with off-pump coronary artery surgery in women. Ann Thorac Surg 2002; 74(6):2113-9.
20. Bucerious J, Gummert JF, Walther T et al. Impact of diabetes mellitus on cardiac surgery outcome. Thorac Cardiovasc Surg 2003; 51(1):11-6.
21. Szabo Z, Hakanson E, Svedjeholm R. Early postoperative outcome and medium-term survival in 540 diabetic and 2239 nondiabetic patients undergoing coronary artery bypass grafting. Ann Thorac Surg 2002; 74(3):712-9.
22. Carson JL, Scholz PM, Chen AY et al. Diabetes mellitus increases short-term mortality and morbidity in patients undergoing coronary artery bypass graft surgery. J Am Coll Cardiol 2002; 40(3):424-7.
23. Society of Thoracic Surgery data base. STSdatabase.www.sts.org.

CHAPTER 9

Understanding the Mechanisms of Hemodynamic and Echocardiographic Changes during OPCAB Surgery

Pierre Couture, André Denault, Patrick Limoges, Peter Sheridan and Denis Babin

Coronary artery bypass grafting (CABG) on the beating heart has become a widely applied procedure. OPCAB grafting is quite attractive because of the obvious advantages of avoiding cardiopulmonary bypass (CPB)-related complications. However, the limitations to this approach have been a higher rate of incomplete revascularization and the threat of intra-operative hemodynamic instability.

The basic principles of complete revascularization are deeply anchored in the cardiovascular community and should not be compromised.[1] In a review of 3,372 surgical patients from the Coronary Artery Surgery Study Registry with triple-vessel disease, those with severe angina (New York Heart Association class III and IV) or left ventricular dysfunction (ejection fraction < 35%) had better 6-year event-free survival when grafts to 3 or more vessels were completed.[2] In the early days of OPCAB surgery, incomplete revascularization with off-pump procedures was frequently reported. Surgeons then would prefer to voluntarily ungraft the circumflex coronary artery (CX) to ease the procedure.[3] Tasdemir et al[4] identified ungrafted CX stenosis as a risk factor for morbidity and mortality, but this has largely been overcome in more recent experience.[5,6]

To obtain adequate exposure and achieve complete revascularization, particularly in patients with difficult lateral CX and posterior branches, midline sternotomy with appropriate methods of mobilization and coronary artery stabilization, remain the most popular approach for multivessel OPCAB.[1,7,8] The use of pericardial sutures to elevate the heart has extended OPCAB application from single left anterior descending (LAD) graft to multivessel coronary surgery.[1] However, heart mobilization and stabilization of the heart during these procedures are often associated with hemodynamic instability.

Various devices are being marketed for the purpose of restricting regional wall motion: the Diamond Grip Rib Spreader Cardiac Stabilizer (Genzyme Corp., Cambridge, MA), the Origin Cardiac Stabilizer and Stabilizer Foot (Origin, Menlo Park, CA), the Mechanical Stabilizer (CTS Inc., Cupertino, CA), the Octopus Tissue Stabilizer System (Octopus, Medtronics, Inc., Minneapolis, MN) and the fork-type compression stabilizer (CorVasc of CoroNéo Inc., Montreal, Quebec, Canada) developed and currently used at the Montreal Heart Institute.[1,9-11] While these devices are different, their stabilizing effects could generally be attributed to a compression or suction effect.[10]

Off Pump Coronary Artery Bypass Surgery, edited by Raymond Cartier. ©2005 Eurekah.com.

Hemodynamic variations in OPCAB may be due to heart mobilization, stabilization, or myocardial ischemia during coronary occlusion. In this chapter, we review and discuss the hemodynamic effects of suction and compression-type stabilizing devices as well as the value and limitations of electrocardiographic, hemodynamic and echocardiographic monitoring modalities during OPCAB.

Hemodynamic Changes after Mobilization and Stabilization

Suction-Type Stabilizer (Octopus, Medtronics Inc.)

Hemodynamic variations occurring during beating heart surgery with a suction-type stabilizer have been studied extensively in animal models. Borst et al[12] from Utrecht, the Netherlands, developed a mechanical suction stabilization system (Octopus, Medtronic Inc.). They demonstrated minimal arrhythmogenesis, hemodynamic deterioration, and histological changes related to suction application of the stabilizing device in pigs. Moreover, they obtained excellent reproducible reduction of cardiac surface motion. Although no hemodynamic deterioration was observed in their model during grafting of the LAD artery and right coronary (RC) artery, the CX territory was more challenging. Exposing posterior branches by displacing the apex of the beating heart decreased arterial pressure in both experimental model[13] and clinical set-up.[14] Experimentally, Grundeman[13] and Jansen et al[14] reported on the feasibility of immobilizing the posterolateral cardiac wall with the straight Octopus paddle fixed directly on the ventricle. The apex was progressively raised anteriorly over a 2-min period by pulling on the left ventricle (LV). This heart "dislocation" caused a 26% decrease in mean arterial pressure (MAP), a 37% fall in cardiac output (CO), and biventricular failure characterized by a major drop in stroke volume (44%) despite an elevation of right ventricular end-diastolic pressure and unchanged left ventricular end-diastolic pressure (Fig. 1). The 20° head-down position (Trendelenburg) normalized CO and MAP. They documented decreased of 34%, 25% and 50% respectively in coronary artery blood flow of the LAD, RC artery, and the CX artery.[15] 20° head-down tilt restored these parameters (Fig. 2). Employing a similar model,[16] these authors further elucidated the mechanism of ventricular dysfunction. They observed a significant compresssion of the right ventricular free wall against the interventricular septum, whereas the right ventricle (RV) outflow tract was narrowed, resulting in a decline of RV dimension with a smaller reduction of LV dimension (Fig. 3). No valvular incompetence was discerned. Trendelenburg positioning normalized MAP, stroke volume and LV dimension, while it partially corrected RV dimension (Fig. 4, study 1). In addition, right heart bypass increased stroke volume and MAP by augmenting LV preload. In contrast, LV bypass failed to restore the systemic circulation (Fig. 4, study 2). They concluded that, in their model, 90° anterior displacement of the apex caused RV deformation and ventricular dysfunction without valvular incompetence or outflow obstruction. These data are in agreement with a similar experiment in a sheep model[17] where RV heart bypass also restored LV function.

This device has also been used successfully for multiple vessel OPCAB in human studies[14,18-20] with minimal morbidity and mortality. Hemodynamic changes have been less characterized in humans. London et al[19] found that gradual vertical positioning of the heart could be accomplished safely in the majority of patients, irrespective of LV function. They bypassed the CX artery in 58% of their cases. Jansen et al[21] have published their results on the first 100 patients with the Octopus stabilizer. They reported that through median sternotomy, exposure of the anterior wall and inferior wall was hemodynamically well-tolerated as long as heart dislocation was performed slowly (approximately over 1 min) (Table 1). However, fluid redistribution (Trendelenburg maneuver) and dopamine were necessary in 67% of patients to maintain arterial pressure with posterior wall exposure. Recently, in a study of 150 patients

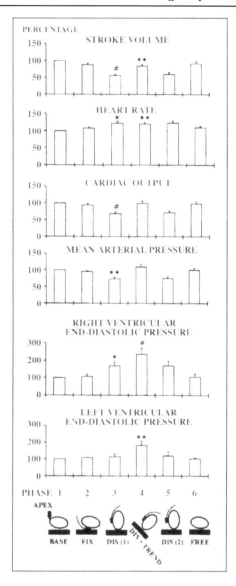

Figure 1. Relative changes in hemodynamic parameters during vertical displacement of the beating porcine heart by the Utrecht Octopus and the effect of head-down tilt (BASE= pericardial control position, FIX= fixation of the suction tentacles to the posterior cardiac wall, DIS (1)= displacement of the heart by the Octopus, TREND= Trendelenburg maneuver (20° head-down tilt), DIS (2)= retracted heart with table returned to horizontal position, and FREE= pericardial position after release of the Octopus). Statistical comparison with control values: * $P < 0.05$, ** $P < 0.001$. Reprinted with permission from the Society of Thoracic Surgeons. Ann Thorac Surg 1997; 63:S90.

Nierich et al.[18] showed that stroke volume was affected mainly during revascularization of the diagonal artery (-25%). They attributed this drop to right heart compression between the LV and the right pericardial wall.

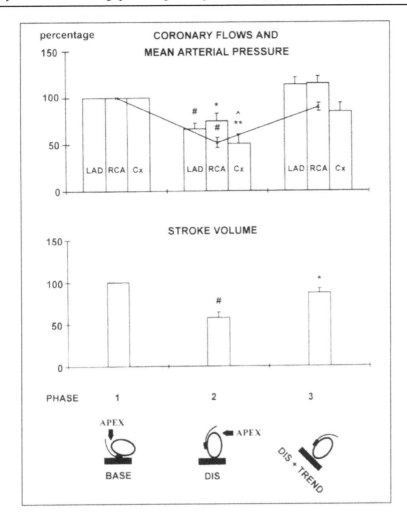

Figure 2. Relative changes in hemodynamic parameters during vertical displacement of the beating porcine heart by the Medtronics Octopus Tissue Stabilizer and the effect of head-down tilt. (BASE= pericardial control position, CX= circumflex coronary artery, DIS= displacement of the heart by the Octopus, DIS + TREND= Trendelenburg maneuver (20° head-down tilt) while the heart remained 90° retracted, LAD= left anterior descending coronary artery, RCA= right coronary artery, X= mean arterial pressure. Statistical comparison with control values * $P < 0.05$, ** $P < 0.01$, # $P < 0.001$, ^ $P < 0.046$ vs combined relative value of LAD and RCA flows. Reprinted with permission from the Society of Thoracic Surgeons. Ann Thorac Surg 1998; 65:1350.

From these observations, it appears that positioning of the heart, particularly during heart dislocation, primarily causes hemodynamic changes. Coronary occlusion during coronary artery anastomosis can further affect LV function, especially if collateral blood flow is not adequate. Using intra-operative transesophageal echocardiography (TEE) in patients undergoing occlusion of the LAD artery lasting up to 15 min during grafting, Koh et al[22] noted LV systolic and diastolic dysfunction in subjects without collaterals. Only diastolic dysfunction was been in patients with collateral flow probably due to the stabilizer itself. These disturbances normalized within 10 min of reperfusion.

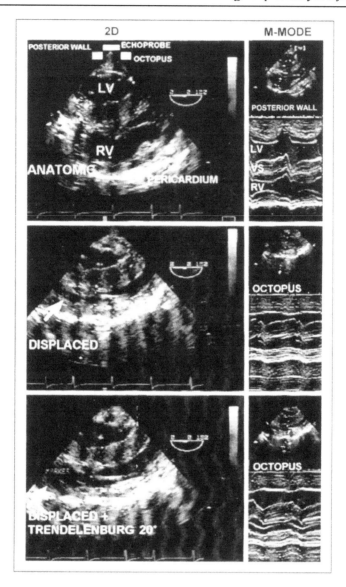

Figure 3. Echographic cross-sectional areas (left panels) and M-mode images (right panels) obtained during 90° anterior displacement (DISPLACED) and subsequent head-down positioning (+TRENDELENBURG). VS= ventricular septum, RV= right ventricle, LV= left ventricle. The ultrasound probe was positioned between two Octopus tentacles and aligned to the posterior wall (ANATOMIC). Note the severe deformation of the RV on cardiac retraction (arrows, DISPLACED) and partial recovery of the RV cross-sectional area after the Trendelenburg maneuver. Reprinted with permission from The Society of Thoracic Surgeons. J Thorac Cardiovasc Surg 1999; 118:317.

Compression-Type Stabilizer

Off-pump multivessel coronary artery surgery began at the Montreal Heart Institute in September 1996.[11] Our experience with beating heart surgery is reported in Chapter 8. Details of the surgical technique have been described elsewhere.[1] All surgeries were performed through

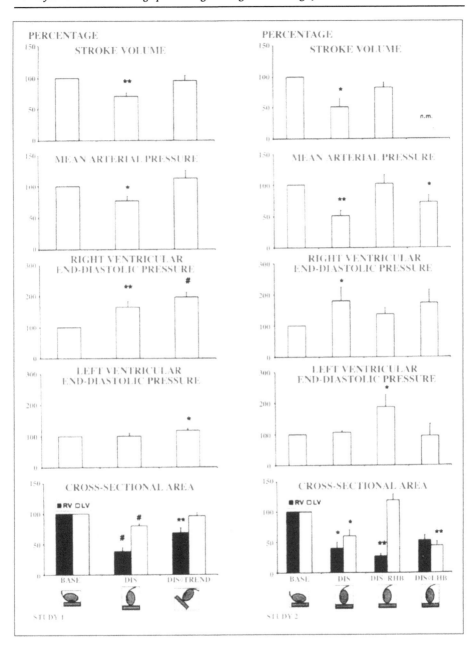

Figure 4. Relative changes in hemodynamic parameters and RV and LV cross-sectional areas during vertical displacement of the beating porcine heart with the Octopus tissue stabilizer (study 1), and the effect of mechanical circulatory support on displacement (study 2). Statistical comparison with control values * P < 0.05, ** P < 0.01, # P < 0.001. n.m.: not measured; erroneous readings from the ultrasound aortic flow probe caused by turbulent flow from the LHB cannula. DIS= displaced, LV= left ventricle, RV= right ventricle, RHB= right heart bypass, LHB= left heart bypass. Reprinted with permission from The Society of Thoracic Surgeons. J Thorac Cardiovasc Surg 1999; 118:318.

Table 1. Hemodynamic changes during immobilization and presentation of the target vessel by the Octopus tissue stabilizer

	Limited Access	Full Access		
	Anterior Wall (n = 49)	Anterior Wall (n = 39)	Inferior Wall (n = 32)	Posterior Wall (n = 12)
Heart rate (beats/min^{-1})				
Before Octopus	66 ± 11	65 ± 12	69 ± 12	66 ± 12
During Octopus	70 ± 11	75 ± 14	74 ± 14	77 ± 11
After Octopus	70 ± 12	72 ± 13	74 ± 13	77 ± 12
Mean Arterial Pressure (mmHg)				
Before Octopus	72 ± 9	74 ± 11	72 ± 11	73 ± 13
During Octopus	70 ± 11	70 ± 12	67 ± 10	66 ± 15
After Octopus	70 ± 11	69 ± 10	68 ± 8	60 ± 15
Cardiac Index (L · min^{-1} m^{-2})				
Before Octopus	3.3 ± 1.1	2.6 ± 0.5	2.9 ± 0.6	2.8 ± 0.6
During Octopus	3.1 ± 0.9	2.5 ± 0.4	2.6 ± 0.6	2.8 ± 0.5
After Octopus	3.1 ± 0.9	2.6 ± 0.4	2.5 ± 0.5	3.1 ± 0.7
Dopamine Required				
During Octopus (%)	7 (14)	9 (23)	10 (31)	8 (67)

Data are means ± 1 standard deviation. Data on the subxiphoid and left posterior access group are not included. "During Octopus" and "after Octopus" data respectively represent measurements 5 min after having completed and "frozen" presentation. Reprinted with permission from the Society of Thoracic Surgeons. J Thorac Cardiovasc Surg 1998; 116:65.

full median sternotomy. The anesthetic technique was left to the discretion of the attending anesthesiologist. Neosynephrine and nitroglycerine were infused as required to normalize hemodynamics. Immobilization of the target artery was achieved through mechanical stabilization. The heart was verticalized to expose the circumflex territory through 4 deep pericardial sutures placed between the left superior pulmonary vein and the inferior vena cava. Extraction of the apex was achieved with minimal distortion since no direct traction was applied on it. Rotation of the surgical table and Trendelenburg positioning assisted in exposing the posterior wall and maintaining RV preload.

Hemodynamic changes associated with this technique have been reported recently.[23] Thirty-one patients undergoing beating heart CABG were studied. They were given no preconditioning. A decrease in mean systemic arterial pressure (SAP) was observed according to the grafted territory: LAD artery: -11 ± 19%; diagonal artery (DA): -13 ± 27%; CX: -19 ± 17%; and posterior descending artery (PDA): -17 ± 14%. Pulmonary artery pressure (PAP) elevation was more significant during DA (+47 ± 84%) than LAD (+30 ± 36%), CX (+21 ± 48%) or PDA (+10 ± 24%) grafting. Changes in mixed venous oxygen saturation (SvO$_2$) were minimal. The authors concluded that myocardial mobilization and stabilization were mainly responsible for hemodynamic variations. The suggested explanation for the striking PAP elevation during DA revascularization was LV outflow tract obstruction caused by the stabilizer compression. To further define the hemodynamic changes associated with this "fork-type" stabilizer, we studied an additional 53 patients undergoing OPCAB surgery in whom we also measured the CO and SvO$_2$.[24] TEE in 5 of these patients assessed the systolic and diastolic functions, to detect regional wall motion abnormalities and mitral regurgitation during the procedure. The LV was divided into 16 segments according to the recommendations of the American Society

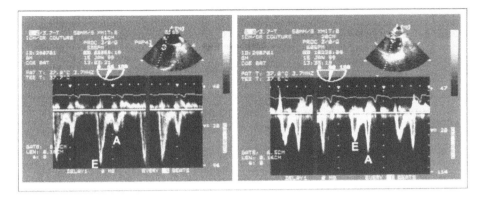

Figure 5. Effect of diagonal artery clamping on mitral E and A waves. Occlusion of the diagonal artery was associated with a reduction of E velocity with and an increase in A velocity compatible with ischemia or preload reduction through compression of the LV outflow tract. Reprinted with permission from Figure 7, Can J Anesth 2002; 49:841.

of Echocardiography.[25] Mitral inflow velocities were measured at the tip of the mitral leaflets on 3 consecutive heartbeats at the end of expiration. The recorded parameters were: peak velocity of the early diastolic filling wave (E), the late diastolic filling wave (A), and the ratio of these 2 velocities (E/A). Our results confirmed those reported earlier.[23] Reduced SAP was observed after stabilization of the LAD (-8.6 ± 22.1%), DA (-14 ± 25.2%), CX (-16.3 ± 13.8%) and RC (-5.1 ± 15.6%) arteries. Antero-posterior compression of the beating heart was responsible for the most significant changes in PAP. PAP elevation was only significant during DA (+37.7 ± 73.6%) and was more pronounced with LAD (+23.9 ± 35.2%) than CX (+12.5 ± 35.9%) or RC revascularization (+12.7 ± 23.8%). Changes in SvO_2 were < 10%, and the decrease in cardiac index was more apparent during DA revascularization (-14.8 ± 16.8%). Cardiac wall motion and mitral valve function were unaffected, even when the heart was displaced at 90° for CX exposure. Our results indicate that the changes in LV mitral inflow were due to LV diastolic dysfunction rather than systolic regional dysfunction.[26] The E/A ratio, which describes such dysfunction, was greatly impaired following DA and LAD stabilizations (Fig. 5). It is inferred that compression on the beating heart during DA and, to a lesser extent, LAD stabilization prevented normal diastolic LV expansion. Applying minimal force on the heart, especially when dealing with the DA, may attenuate such a problem.

Only one other study evaluated cardiac function during beating heart CABG with a compression-type stabilizer and invasive hemodynamic as well as TEE monitoring. In this study, Jurmann et al[27] assessed 28 patients undergoing single bypass grafting to the LAD artery with the internal thoracic artery through a small left anterior thoracotomy. They used a compression-type of stabilizer and applied gentle pressure until sufficient immobilization of the anastomotic site was achieved at the level of the LAD artery. They reported that application of the epicardial stabilizer resulted in a minor decrease of left ventricular end-diastolic and systolic diameter, with unchanged fractional area, cardiac index, stroke volume index and pulmonary capillary wedge pressure (PCWP). In contrast to our findings, however, after 10 min of occlusion and with the stabilizer in place, left ventricular systolic and diastolic disease increased in comparison to values obtained just after placement of the stabilizer; the cardiac index and stroke volume index declined moderately (Table 2). This was associated with compromised LV regional wall motion characterized by new hypokinetic segments or the deterioration of preexisting hypokinetic segments. In general, this compromise in LV function did not reach critical levels, and was easily compensated by appropriate pharmacological intervention. These different observations, compared to our results, may be attributed to operative technique,

Table 2. *Left ventricular dimensions and function from intra-operative transesophageal echocardiography during invasive direct coronary artery bypass grafting*[a]

Variable	Baseline	Stabilizer	Ischemia	Reperfusion
FAC (%)	59.9 ± 15.6	61.4 ± 18.1	56.9 ± 17.9[b]	65.8 ± 15.3[c]
LVEDD a-p (mm)	38.4 ± 7.3	35.6 ± 8.7[b]	37.2 ± 8.2[b]	37.5 ± 6.9
LVESD a-p (mm)	27.2 ± 7.7	25.2 ± 7.8[b]	26.8 ± 7.8[b]	24.8 ± 6.8[c]
LVEDD s-l (mm)	38.6 ± 7.5	37.3 ± 7.6[b]	38.0 ± 7.7	38.1 ± 8.2
LVESD s-l (mm)	28.6 ± 7.5	26.9 ± 8.6[b]	28.6 ± 9.6[b]	27.3 ± 8.1
LVEDD apex (mm)	24.6 ± 3.7	21.6 ± 5.1[b]	24.6 ± 4.2[b]	25.6 ± 3.4
LVESD apex (mm)	11.2 ± 7.5	10.5 ± 7.8[b]	14.4 ± 4.9[b]	10.2 ± 8.3

[a] Data are means ± 1 standard deviation; [b] Significance: *P* < 0.05 *vs* previous condition; [c] Significance: *P* < 0.05 *vs* baseline. a-p= anterior-posterior plane, s-l= septal-lateral plane, FAC= fractional area change, LVEDD= left ventricular end-diastolic diameter, LVESD= left ventricular end-systolic diameter. Reprinted with permission from the Society of Thoracic Surgeons. Ann Thorac Surg 1998; 66:1084.

type of stabilizing device, and patient characteristics. Our patients had multivessel disease, with different collateral networks, which are known to affect LV function during vessel occlusion.[22] Our surgical strategy of revascularizing the most stenotic vessel first (then the most collateralized) could have been a significant and determining factor too.[1] Although the different types of stabilizers (compression vs. suction) were not specifically compared, the current literature suggests that stabilizer type does not influence hemodynamic changes.[28]

Monitoring Myocardial Ischemia

Intra-operative monitoring of myocardial ischemia during CABG is generally accomplished by electrocardiography with the combination of lead 2 and V5 for ST segment changes (Fig. 6).[19] Mobilization of the heart, particularly during off-pump grafting of the CX territory, often causes microvoltages on EKG. The value of electrocardiography and ST segment monitoring under these circumstances is less predictable. Verticalization-related micro-voltage (the heart being "out" of the chest cavity) reduces the sensitivity of ST segment monitoring, making EKG unreliable for surveillance of ischemia. TEE is a recognized, sensitive detection method for myocardial ischemia during conventional CABG surgery through evaluation of segmental wall motion abnormalities (SWMA).[29,30] However, very few studies have explored the role of TEE during OPCAB surgery. Investigationg 27 patients undergoing OPCAB surgery, Moises et al[31] monitored SWMA with TEE during coronary artery grafting. No stabilizers were used. A total of 48 anastomosis, 26 to the LAD, 3 to the DA, and 19 to the RC, were performed. They documented 31 (64%) new SWMA during coronary occlusion as defined by the American Society of Echocardiography.[25] For each segment, a score was assigned as follows: normal = 1; hypokinetic = 2, akinetic = 3, and dyskinetic = 4. By the time the chest was closed, complete recovery had occurred in 16 (50%) segments, partial recovery in 10 (33%), and no recovery in 5 (17%). Electrocardiographic ST-T changes suggestive of myocardial ischemia occurred during only 9/48 anastomoses, confirming the better sensitivity of TEE in detecting intra-operative myocardial ischemia. On the 7th post-operative day, new SWMA persisted in 5 segments that had not recovered at the end of the surgery and in 2 (20%) of 10 segments with partial recovery, reinforcing the value of TEE in predicting persistent SWMA after revascularization. This observation is quite pertinent because detection of persistent SWMA is associated with increased cardiac enzyme levels, and increased post-operative morbidity, such as pulmonary edema

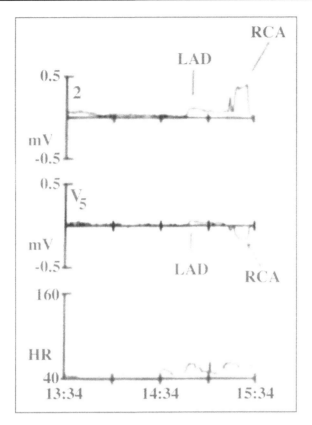

Figure 6. Ischemic changes during off-pump bypass cardiac surgery. Significant ST segment elevations were observed during left anterior descending (LAD) artery and right coronary artery (RCA) occlusion. Reprinted with permission from Figure 8, Can J Anesth 2002; 49:843.

as well as atrial fibrillation.[31,32] The presence of persistent SWMA after revascularization should lead to coronary bypass graft revision.

In another study, Kotoh et al[33] undertook colour kinesis TEE to evaluate regional wall motion. Colour kinesis measured endocardial mobility by acoustic quantification and wall motion changes on a single end-systolic frame (Fig. 7). In 34 beating heart CABG patients, SWMAs were observed in 4 patients (12%) undergoing left internal mammary artery to LAD artery grafting. Only 1 patient developed electrocardiographic abnormalities. The pericardial tractions applied during CX and RC artery revascularization occasionally impeded echocardiographic views of the LV. On these occasions, TEE may be insufficient to evaluate wall motion suggestive of myocardial ischemia. While monitoring of PAP along with PCWP has been shown to be an insensitive marker of myocardial ischemia during CABG surgery under CPB,[30,33] it remains the only reliable method of ischemia-monitoring during displacement of the heart during OPCAB surgery. Decreased ventricular compliance during ischemia increases filling pressure, ultimately heightening PAP and PCWP. VanDaele et al[34] recorded 33% sensitivity for a PCWP increase of 3 mmHg over baseline to detect myocardial ischemia compared to abnormal wall motion found by TEE in patients under general anesthesia. Patients who experienced elevated PCWP produced documentation of inferior wall ischemia with transient papillary muscle dysfunction and some degree of mitral valve regurgitation, explaining the PAP rise. According to these observations, it is quite obvious that all methods of

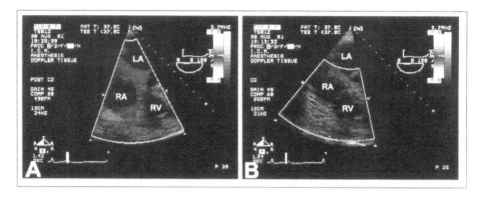

Figure 7. Tissue colour Doppler illustrating a reduction in colour-coded intensity at the basal region of the right ventricle with occlusion of the right coronary artery during off-pump cardiac bypass surgery. A) Baseline. B) After right coronary artery stabilization and clamping.

monitoring are not exclusive but rather complementary. Surgeons as well as anesthesiologists must rely on combinations of these methods to optimize the detection of myocardial ischemia during OPCAB surgery.

At the Montreal Heart Institute, TEE is generally undertaken for noncoronary surgery, although it has been more frequently employed during OPCAB surgery lately. Our experience with TEE during cardiac surgery has been published.[35] TEE influenced the course of coronary artery bypass in 10% of our patients. We found it to be most helpful during cardiac manipulations associated with OPCAB surgery. In these situations, TEE helps to differentiate between ischemic dysfunction and increased filling pressure due to extra-cardiac compression (Fig. 8). TEE is also useful to detect unexpected ischemic mitral regurgitation, which could occasionally lead to mitral valve repair or replacement instead of revascularization alone. Our group has recently underscored the importance of diastolic function evaluation during cardiac surgery.[36]

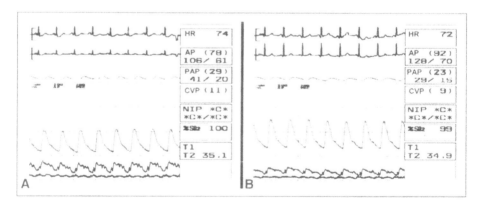

Figure 8. A) Off-pump cardiac bypass surgery associated with elevated central venous pressure and pulmonary artery pressure followed by hypotension secondary to compression of the left ventricle during left anterior descending coronary artery anastomosis. This was not associated with any regional wall motion abnormalities, but was secondary to extracardiac compression. B) Effect of removal of the "fork type" stabilizer positioned over the left anterior descending artery on hemodynamic variables. An increase in systemic blood pressure, a reduction in pulmonary artery pressure and in central venous pressure resulted in removal of the stabilizer. Reprinted with permission from Figure 9, Can J Anesth 2002; 49:844.

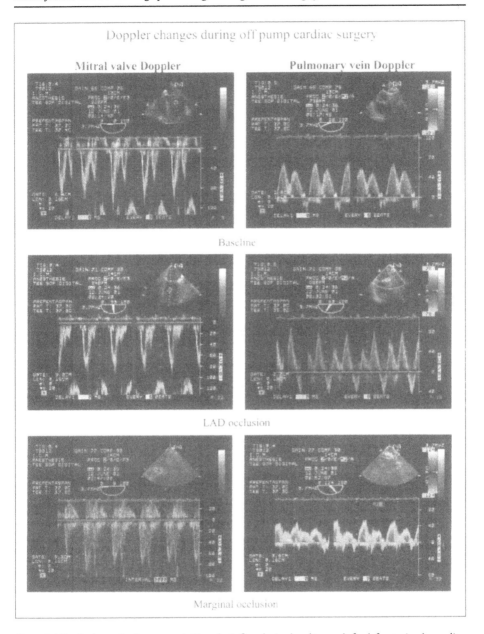

Figure 9. Mitral valve and pulmonary vein Doppler inflow during baseline, and after left anterior descending (LAD) and marginal artery occlusion. During LAD occlusion, a restrictive pattern appears with a predominant E wave and a reduced S to D wave on the pulmonary vein Doppler signal. This anomaly does not occur during marginal artery clamping. Reprinted with permission from Figure 10, Can J Anesth 2002; 49:845.

We currently use Doppler to evaluate both left (Fig. 9) and right diastolic function (Fig. 10) during off-bypass surgery. In our hands, such evaluation appears to be quite promising in improving our understanding of hemodynamic changes during beating heart surgery.

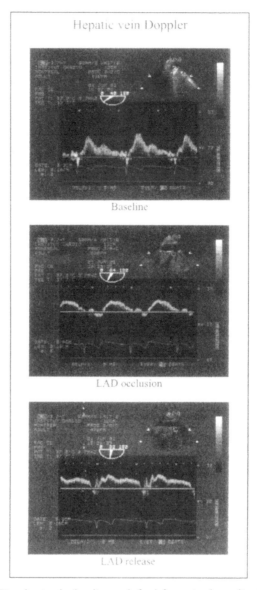

Figure 10. Hepatic vein Doppler signal at baseline, and after left anterior descending (LAD) occlusion and release. During LAD occlusion, a reversible restrictive pattern on the hepatic vein Doppler signal appears with a reversal of the S wave signal. Reprinted with permission from Figure 11, Can J Anesth 2002; 49:845.

In summary, hemodynamic changes represent an important diagnostic challenge during off-bypass procedures. The mechanisms vary according to the surgical technique used. EKGs and classic hemodynamic monitoring devices remain limited in establishing the cause of hemodynamic instability. Intra-operative TEE can help to pinpoint the etiology of hemodynamic instability and reorient the surgical therapy. The impact and cost-effectiveness of routine TEE during off-bypass CABG have not yet been evaluated, but appear promising in assisting surgeons, during OPCAB procedures.

References

1. Cartier R, Blain R. Off-pump revascularization of the circumflex artery: Technical aspect and short-term results. Ann Thorac Surg 1999; 68:94-9.

2. Bell MR, Gersh BJ, Schaff HV et al. Effect of completeness of revascularization on long-term outcome of patients with three-vessel disease undergoing coronary artery bypass surgery. A report from the Coronary Artery Surgery Study (CASS) Registry. Circulation 1992; 86:446-57.

3. Gundry SR, Romano MA, Shattuck OH et al. Seven-year follow-up of coronary artery bypasses performed with and without cardiopulmonary bypass. J Thorac Cardiovasc Surg 1998; 115:1273-7.

4. Tasdemir O, Vural KM, Karagoz H et al. Coronary artery bypass grafting on the beating heart without the use of extracorporeal circulation: Review of 2052 cases. J Thorac Cardiovasc Surg 1998; 116:68-73.

5. Puskas JD, Williams WH, Duke PG et al. Off-pump coronary artery bypass grafting provides complete revascularization with reduced myocardial injury, transfusion requirements, and length of stay: A prospective randomized comparison of two hundred unselected patients undergoing off-pump versus conventional coronary artery bypass grafting. J Thorac Cardiovasc Surg 2003; 125(4):797-808.

6. Cartier R. Current trends and technique in OPCAB surgery. J Card Surg 2003; 18(1):32-46.

7. Benetti FJ, Naselli G, Wood M et al. Direct myocardial revascularization without extracorporeal circulation. Experience in 700 patients. Chest 1991; 100:312-6.

8. Shennib H, Mack MJ, Lee AG. A survey on minimally invasive coronary artery bypass grafting. Ann Thorac Surg 1997; 64:110-4.

9. Cartier R, Hébert Y, Blain R et al. Triple coronary artery revascularization on the stabilized beating heart: Initial experience. Can J Surg 1998; 41:283-8.

10. Dagenais F, Perrault LP, Cartier R et al. Beating heart coronary artery bypass grafting: Technical aspects and results in 200 patients. Can J Cardiol 1999; 15:867-72.

11. Cartier R. From idea to operating room: Surgical innovation, clinical application, and outcome. Sem Cardiothoracic Vascular Anesthesia 2000; 2:103-9.

12. Borst C, Jansen EW, Tulleken CA et al. Coronary artery bypass grafting without cardiopulmonary bypass and without interruption of native coronary flow using a novel anastomosis site restraining device ("Octopus"). J Am Coll Cardiol 1996; 27:1356-64.

13. Grundeman PF, Borst C, van Herwaarden JA et al. Hemodynamic changes during displacement of the beating heart by the Utrecht Octopus method. Ann Thorac Surg 1997; 63:S88-S92.

14. Jansen EW, Grundeman PF, Mansvelt Beck HJ et al. Experimental off pump grafting of a circumflex branch via sternotomy using a suction device. Ann Thorac Surg 1997; 63:S93-S6.

15. Grundeman PF, Borst C, van Herwaarden JA et al. Vertical displacement of the beating heart by the octopus tissue stabilizer: Influence on coronary flow. Ann Thorac Surg 1998; 65:1348-52.

16. Grundeman PF, Borst C, Verlaan CW et al. Exposure of circumflex branches in the tilted, beating porcine heart: Echocardiographic evidence of right ventricular deformation and the effect of right or left heart bypass. J Thorac Cardiovasc Surg 1999; 118:316-23.

17. Porat E, Sharony R, Ivry S et al. Hemodynamic changes and right heart support during vertical displacement of the beating heart. Ann Thorac Surg 2000; 69:1188-91.

18. Spooner TH, Dyrud PE, Monson BK et al. Coronary artery bypass on the beating heart with the Octopus: A North American experience. Ann Thorac Surg 1998; 66:1032-35.

19. London MJ, Hollenberg M, Wong MG et al. Intraoperative myocardial ischemia: Localization by continuous 12-lead electrocardiography. Anesthesiology 1988; 69:232-41.

20. Nierich AP, Diephuis J, Jansen EWL et al. Heart displacement during off-pump CABG: How well is it tolerated? Ann Thorac Surg 2000; 70:466-72.

21. Jansen EW, Borst C, Lahpor JR et al. Coronary artery bypass grafting without cardiopulmonary bypass using the octopus method: Results in the first one hundred patients. J Thorac Cardiovasc Surg 1998; 116:60-7.

22. Koh TW, Carr-White GS, DeSouza AC et al. Effect of coronary occlusion on left ventricular function with and without collateral supply during beating heart coronary artery surgery. Heart 1999; 81:285-91.

23. Do QB, Cartier R. Hemodynamic changes during bypass surgeries in the beating heart. Ann Chir 1999; 53:706-11.

24. Do QB, Chavanon O, Couture P et al. Hemodynamic repercussion during beating-heart CABG surgery. Can J Cardiol 1999; 15:267.

25. Schiller NB, Shah PM, Crawford M et al. Recommendations for quantitation of the left ventricle by two-dimensional echocardiography. American Society of Echocardiography Committee on Standards, Subcommittee on Quantitation of Two-Dimensional Echocardiograms. J Am Soc Echocardiogr 1989; 2:358-67.

26. Rakowski H, Appleton C, Chan KL et al. Canadian consensus recommendations for the measurement and reporting of diastolic dysfunction by echocardiography: From the investigators of consensus on diastolic dysfunction by echocardiography. J Am Soc Echocardiogr 1996; 9:736-60.

27. Jurmann MJ, Menon AK, Haeberle L et al. Left ventricular geometry and cardiac function during minimally invasive coronary artery bypass grafting. Ann Thorac Surg 1998; 66:1082-6.

28. Gummert JF, Raumanns J, Bossert T et al. Suction device versus pericardial retraction sutures. Comparison of hemodynamics using different exposure systems in beating heart coronary surgery. Heart Surg Forum 2003; 6(Suppl 1):S32.

29. Smith JS, Cahalan MK, Benefiel DJ et al. Intraoperative detection of myocardial ischemia in high-risk patients: Electrocardiography versus two-dimensional transesophageal echocardiography. Circulation 1985; 72:1015-21.

30. Leung JM, O'Kelly B, Browner WS et al. Prognostic importance of postbypass regional wall-motion abnormalities in patients undergoing coronary artery bypass graft surgery. SPI Research Group. Anesthesiology 1989; 71:16-25.

31. Moises VA, Mesquita CB, Campos O et al. Importance of intraoperative transesophageal echocardiography during coronary artery surgery without cardiopulmonary bypass. J Am Soc Echocardiogr 1998; 11:1139-44.

32. Malkowski MJ, Kramer CM, Parvizi ST et al. Transient ischemia does not limit subsequent ischemic regional dysfunction in humans: A transesophageal echocardiographic study during minimally invasive coronary artery bypass surgery. J Am Coll Cardiol 1998; 31:1035-9.

33. Kotoh K, Watanabe G, Ueyama K et al. On-line assessment of regional ventricular wall motion by transesophageal echocardiography with color kinesis during minimally invasive coronary artery bypass grafting. J Thorac Cardiovasc Surg 1999; 117:912-7.

34. van Daele ME, Sutherland GR, Mitchell MM et al. Do changes in pulmonary capillary wedge pressure adequately reflect myocardial ischemia during anesthesia? A correlative preoperative hemodynamic, electrocardiographic, and transesophageal echocardiographic study. Circulation 1990; 81:865-71.

35. Couture P, Denault AY, McKenty S et al. Impact of routine use of intraoperative transesophageal echocardiography during cardiac surgery. Can J Anaesth 2000; 47:20-6.

36. Bernard F, Denault A, Goyer C et al. Diastolic dysfunction is of additional value in predicting difficult weaning from cardiopulmonary bypass. Anesth Analg 2001; 92:291-8.

CHAPTER 10

Cerebral Complications Following Coronary Artery Bypass Grafting Surgery

Marzia Leache and Raymond Cartier

Cerebral complications constitute the leading cause of morbidity and disability after heart surgery. Although many cerebral deficits resolve with time, others remain a major handicap with devastating effects on both patients and their families. The reported incidence of perioperative stroke ranges from 0.4% to 5.4% and from 25% to 79% for neuropsychological dysfunction.[1-4] This variability depends on a number of factors, some related to the patients themselves, and others to the type of study (retrospective or prospective) or evaluation tools. With more than 800,000 coronary artery bypass grafting (CABG) procedures being performed annually worldwide, the social and economic implications of these complications are quite significant.

Classification Based on Cerebral Outcomes

Four different types of neurological and cognitive complications have been observed after CABG surgery: stroke, post-operative delirium, short-term, and long-term neurocognitive changes. Based on the different etiologies, neurological complications can be classified into 2 types of cerebral outcomes: type I includes death by stroke or hypoxic encephalopathy, nonfatal stroke, transient ischemic attack (TIA), stupor, seizure, or coma, and type II includes milder deficits, such as new deterioration of intellectual function, confusion, agitation, disorientation, and memory deficit.[5,6] Type I outcomes are related to macroemboli occluding flow in arteries of 200 μm in diameter or greater and also to severe peri-operative hypotension or prolonged circulatory arrest. These macroemboli originate mostly from atherosclerotic plaques (ascending aorta), air, fat or platelet aggregates; they normally cause focal or multi-focal brain deficits. Severe hypotension or hypoxia can lead to diffuse brain damage mostly affecting the neocortex and the cerebellum. Type II outcomes are associated with microembolisms impairing blood flow in the smaller arteries, arterioles, and capillaries. Air microembolisms from bubble oxygenators, cardiotomy reservoir cellular aggregates, tubing fragments, and silicone antifoam materials are the most common causes. Ultimately, these microembolisms will affect cerebral blood flow without any major, detectable brain damage.[1,2,7-9]

Stroke

The majority of strokes occurring during CABG are caused by cerebral macroemboli. Atheroembolism from the diseased ascending aorta is the most common pathogenic mechanism.[5] In an autopsy study, Blauth and coworkers identified atheroemboli disease in 20% of 221 patients, with most of them being localized in the brain.[10] The majority of affected patients—46 of 48 (95.8%)—presented with severe atherosclerosis of the aorta. Only 2% of the patients had no significant aortic disease. Embolization of atherosclerotic plaques may occur during aortic cannulation, aortic partial or full cross-clamping, and construction of proximal anastomoses.[11] Alternatively, it may also result from the "sandblasting-like" effect caused by the streaming of blood perfusates from the aortic cannula on the diseased aortic wall.[12] Atherosclerosis of the ascending aorta is strongly associated with advanced age. Its prevalence in patients 70 years of age and older reached 34% in certain series.[13] As a result, convincing evidence suggests that advanced age and a history of previous cerebrovascular accidents represent the most important preoperative predictors of perioperative stroke.[14] In a prospective study, Tuman and colleagues reported an exponential increase of stroke with aging during CABG surgery.[15] They recorded an incidence of 0.9% in patients less than 65 years old, 3.6% in patients aged 65 to 74 years, and 9% in patients over 75 years.

Recognition of the diseased aorta based merely on its palpation, chest radiography, or angiography may be inadequate. Systematic epiaortic scanning has emerged lately as the method of choice for the detection of aortic artherosclerosis.[16] Wareing and colleagues using routine epiaortic scanning on 500 patients undergoing CABG, identified 68 (13.6%) with significant atheromatous disease of the aorta.[17] Palpation of the aorta alone revealed atheromatous disease only in 26 of these patients, meaning that palpation alone missed more than 50% of significant arteriosclerosis. Routine epiaortic scanning has been proposed to detect disease of the aorta and decrease the risk of embolization by consequently selecting the optimal cannulation site and the optimal surgical technique.[18] Trans-esophageal echocardiography has also been used to characterize the status of the ascending aorta and the morphology of atheromatous plaque. Pedunculate and noncalcified plaques contain loose, necrotic material and have a higher probability of embolization so that their identification prior to aortic manipulation is a valuable strategy.[19]

Other identified predictors of stroke are: diabetes mellitus, hypertension, severe peri-operative hypotension, left ventricle dysfunction, peripheral vascular disease, second-time surgery, and carotid disease.[6,20,21] The relationship between carotid bruits and stroke is controversial. In a case-control study, Reed and colleagues observed that the presence of carotid bruits increased the risk of stroke or TIA in coronary surgery by 3.9-fold.[22] Conversely, Breuer and Roper, in 2 prospective studies failed to clearly demonstrate a relationship between stroke and carotid bruits.[23,24] The best management of these high-risk patients is far from being established. Discordant data are reported in the literature for the risk of stroke and mortality from various strategies adopted.[25-27] Carotid endarterectomy has been proved superior to medical treatment for symptomatic patients with carotid stenosis greater than 70%.[28] In nonsymptomatic patients undergoing CABG, the timing of the surgery (previous carotid artery endarterectomy before CABG; combined approach; or performing only CABG) is still subject to debate.

Atrial fibrillation and flutter are common arrhythmias following myocardial revascularization, ranging from 15% to 63%.[29,30] Post-operative atrial fibrillation increases resource utilization, prolongs hospital stay, and triples the risk of post-operative stroke.[30] Efforts to recognize patients at risk for such complications, have yielded conflicting results. Only advanced age has been associated with an heightened risk of atrial fibrillation, varying from 4% in patients aged less than 40 years to 50% in patients aged 70 years and older.[31] Clinical trials have focused on prevention with prophylactic utilization of digoxin, magnesium sulphate, β-blocker agents, and amiodarone but no consensus regarding the most effective preventive treatment has yet been reached.[32-37] Anticoagulation (heparin-warfarin) is the most common prophylactic measure used to prevent stroke after the onset of atrial fibrillation.[38,39]

Delirium

Increasing attention has been given to diffuse encephalopathy following cardiac surgery. Several clinical forms have been described: delirium, confusion, memory deficits, cognitive dysfunction, and altered personality or mental state.[40] Diffuse forms of encephalophathy are particularly difficult to distinguish and are frequently ignored due to the short duration of the clinical symptoms and signs that usually resolve within days or weeks. According to Trzepacz, delirium results from a combination of various structural and physiological lesions with primary involvement of the prefrontal cortices, anterior and right thalamus, and right basilar mesial temporoparietal cortex.[41] This common neuroanatomical pathway may be responsible for certain "core symptoms" (disorientation, cognitive deficits, sleep-wake cycle disturbance, disorganized thinking, and language abnormalities), whereas other symptoms (delusions, hallucinations, illusions and affective lability) may depend on the etiology of the delirium. An imbalance in the cholinergic and dopaminergic neurotransmitter system as well as amino acid disturbances are commonly implicated.[42] Post-operative delirium occurs in 10 to 30% of patients after CABG, and is associated with longer hospital stays.[43-45] It is more frequent among the elderly population who already have an augmented risk of functional decline and dementia.[46-49] Rolfson et al have demonstrated that a previous stroke could increase by 8-fold the risk of delirium during conventional cardiac surgery. The duration of cardiopulmonary bypass (CPB) doubles the risk after 38 minutes and triples it after 60.[50] This is the direct consequence of the increase in embolic load that is related to the length of CPB.[51] Other predictors are old age, preoperative nutritional status, history of alcohol abuse and depression as well as structural brain disease.[42,50,52] Significant precipitating factors include infectious and metabolic insults, peripheral vascular disease, substance withdrawal, impaired cerebral oxygenation and perfusion due to low cardiac output syndrome, and pulmonary disease requiring prolonged mechanical ventilation with possible carbon dioxide retension.[5,50,53,54] While a number of predictors for delirium have been identified, their relative roles in the physiopathology of the disorder are still unclear.

Neurocognitive Dysfunction

The reported incidence of neuropsychological dysfunction after CABG ranges from 33% to 83%.[55,56] This variability reflects methodological issues such as the criteria for patient selection, tests used to assess cognition, methods of evaluating test results, the choice of cognitive outcome measurements, and the timing of assessment. There is, however, considerable variation in the prevalence and duration of the deficits.

Short-Term Changes

Neurocognitive impairment starts declining in the first 2 weeks after surgery, but lingers in 30% to 80% of patients.[57,58] At 6 months, it ranges from 24% to 57%, and can persist in up to 35% of patients at 12 months.[59,60] Memory capacity is affected in nearly 30% of patients at 6 months; attention, concentration, motor and mental speed are involved less frequently.[59]

Long-Term Changes

Long-term complaints are subtler and less investigated. The patient may have difficulty with following directions, planning complex actions or making calculations; generally reporting that "they just are not the same." In a longitudinal study, Newman et al found that at the time of discharge, 53% of their 261 patients had declined from presurgical baseline value performance.[61] This was reduced to 36% at 6 weeks and to 24% at 6 months, but was subsequently augmented at 5 years to 42 %. A closer look at their data revealed that only patients with an early decline after surgery were affected by a new decline at 5 years. Selnes and coworkers found a similar biphasic course of cognitive change after cardiac surgery.[62] They demonstrated that performance from baseline to 1 year tended to improve for most cognitive domains; by contrast, from 1 to 5 years, performance declined for most cognitive domains. However, at 5 years, only visuoconstruction and psychomotor speed performances were worse than baseline values. Therefore, the delayed effect may reflect damage and possibly continuing damage to an area of the posterior parietal cortex, called the "watershed area", which is particularly susceptible to hypoperfusion or microembolisms.[63] Moreover, blood flow studies have documented parietal lobe hypoperfusion after myocardial revascularization in patients who sustained neurological complications after CABG and have uncovered a high incidence of posterior watershed area infarcts among them.[64-66] Thus, some of the cognitive changes may be markers of underlying brain changes.

Predictive Factors

Factors associated with short-term outcomes (1 month) are not necessarily associated with long-term (1 year) changes.[67] Advancing age and low educational levels are risk factors for both types of decline.[61,68] Increased embolic load measured intra-operatively by trans-cranial ultrasonography, preoperative high blood pressure, and elevated blood urea nitrogen and creatinine within 7 days post-operatively are linked with short-term cognitive changes.[67,68] Diabetes, severe atherosclerosis of the aorta, a history of stroke, and longer awake time are associated with a long-term decline in specific cognitive domains.[67,68]

Mechanisms of Injury

Two major mechanisms have been implicated in the genesis of cognitive decline after CABG: intra-operative hypotension and multiple microemboli. Mean arterial pressure (MAP) during CPB is usually 50 mmHg. Newman et al have demonstrated a relationship between a MAP value of less than 50 mmHg, age and decreased in spatial and figural memory, suggesting that the combination of hypotension and hypoperfusion is important in older patients.[68] Moody et al in an autopsy study, documented small capillary arteriolar dilatations (SCADs) in the brains of patients who had undergone CABG.[69] They related SCADs to traces of embolic material, mainly lipid, but also silicone and aluminium derived from the CPB apparatus itself. SCAD density appear to increase with the length of bypass time and the return of shed blood to the CPB circuit, as reported previously.[70,71] Other studies have confirmed the association between embolic load and early cognitive decline.[9,72] Factors, such as the use of CPB and its duration, body temperature during CPB, and manipulation of an atheromatous aorta, have been implicated in long-term impairment.[73-76] Tardiff et al also have suggested that genotype is an important predictor of cognitive dysfunction after CPB.[77] They demonstrated a relationship between apolipoprotein E-ε4, a known risk factor for the late onset of Alzheimer's disease, and the decline in cognitive performance on a specific test at the hospital and 6 weeks post-operatively. Thus, impaired, genetically-determined neuronal mechanisms of maintenance and repair may increase the susceptibility of certain individuals to cognitive dysfunction after a physiological stress.

Other Factors Influencing Neurological Outcome after CABG Surgery

Depression

Increasing attention has focused on the effect of depression after myocardial infarction (MI) and heart surgery. Depression increases mortality by 4-fold after MI.[78] In cardiac surgery, its prevalence has reached up to 25% in certain series.[79] Depressed preoperative mood is a major risk factor for developing post-operative depression, but does not correlate with cognitive decline after CABG surgery.[78-80]

The Inflammatory Cascade

The whole body inflammatory reaction during CPB involves complement-activation by foreign surfaces encountered by the blood. This triggers activation of the coagulation pathways, the fibrinolytic and kallikrein cascade, neutrophils, oxygen-derived free radicals and the synthesis of various cytokines.[81-83] These vasoactive substances cause endothelial injury, increased capillary permeability, and accumulation of interstitial fluid. Consequently, general organ dysfunction ensues, including the brain. Cerebral cell impairment is reflected by brain swelling, normally detectable on magnetic resonance imaging assessment within 1 to 2 hours after CABG.[84]

Normothermic and Hypothermic CPB

Cerebral metabolic O_2 consumption correlates with brain temperature. A drop of 10°C in body temperature results in a 68% fall in cerebral oxygen consumption. Hence, hypothermia, by reducing O_2 consumption, confers some cerebral protection against anoxia. Conversely, normothermic CPB should offer better hemodynamic stability, less acidosis, and less alteration of the coagulation cascade, but could expose the patients to a potential incremental risk factor for adverse cerebral outcome. There is conflicting evidence on this issue. The Warm Heart Surgery Investigators assessed 1,732 patients in a randomized trial comparing normothermic vs. hypothermic CPB.[85] They found a similar rate of stroke in both groups (1.6% and 1.5%, respectively). In a comparable study carried out on 1,001 patients, Martin et al reported a greater incidence of stroke (3.1% versus 2.1%) in the normothermic group (body temperature >35°C) compared to the hypothermic group (body temperature ≥ 28°C).[86] These conflicting results may be related to the adverse effects of hypothermia, which protects from cerebral hypoxic or ischemic injury, but could also induce damage by enhancing physiological disturbances associated with the systemic inflammatory response to CPB. Moreover, the need for rewarming before separation from CPB is associated with jugular venous desaturation and widened cerebral arteriovenous oxygen difference.[87] Mild hypothermia (32°C to 35°C) appears to be the current clinical compromise adopted in most centres.

pH Management

During CPB, the 2 acid-base protocols used are alpha-stat and pH-stat.[88,89] During pH-stat management, $PaCO_2$ is maintained constant during hypothermic CPB, and temperature corrected pH of 7.4 is achieved by the addition of CO_2. Consequently, cerebral blood flows, depends on cerebral perfusion pressure or MAP due to the loss of cerebral blood flow autoregulation. The metabolic needs of the brain are fulfilled as long as cerebral perfusion pressure is maintained between 50 to 130 mmHg. Alpha-stat management maintains temperatureuncorrected $PaCO_2$ constant, and allows blood pH to rise in relation to temperature; thus, at temperatures greater than 26°C, autoregulation is preserved until MAP falls below 30 mmHg. There is some evidence of the beneficial effects of alpha-stat management on neurological dysfunction compared to the pH-stat strategy.[90-93] These outcomes are related to the preservation of cerebral autoregulation and lower cerebral blood flow in alpha-stat as opposed to increased flow to the brain, which exceeds metabolic needs with possibly greater embolic load delivered to the brain in pH-stat.

Cerebral Blood Flow during CPB

A vast number of factors influence cerebral blood flow and cerebral metabolism during CPB. These factors include pH strategy, timing of low flow or arrest during hypothermia, flow rates, pulsatile or nonpulsatile flow, blood pressure, intra-cranial pressure, central venous pressure, and hypothermic or normothermic CPB. The relationship between MAP during CPB and cerebral damage is still controversial. Many studies have advocated MAP greater than 50 mmHg to prevent cerebral injury,[94,95] but this has been refuted by others.[96-98] None has correlated lower perfusion pressure levels during CPB with a higher incidence of neurological dysfunction. The same controversy exists regarding pulsatile versus nonpulsatile blood flow during CPB. In experimental studies, pulsatile cerebral perfusion seems to improve gray matter to white matter blood flow ratios and cerebral oxygen consumption.[90] Nonpulsatile perfusion promotes stasis of cerebral fluid, resulting in cerebral edema.[84,99] However, when transposed to clinical outcomes, neither mode of perfusion has demonstrated significant improvement of neurocognitive assessment.[65]

Embolism during CPB

Cerebral embolism is strongly implicated in cerebral damage associated with CPB. Transcranial and carotid doppler scanning have allowed surgeons to investigate the origin and quantity of embolisms during surgery. Macroemboli are likely to arise from surgical manipulation of the heart and the aorta; atheromatous disease of the ascending aorta is the most important causative agent. The majority of microembolisms arise from the extracorporeal circuit or perfusion manipulations.[100] Although the first type usually can cause severe neurological deficits, the latter is more subtle. Clinical manifestations are generally detectable only if multiple embolisms occur, except in the case of susceptible tissue such as the retina. There is a direct correlation between embolic load and the duration of CPB, with a 90% increase for each additional hour of CPB.[51] These embolisms are reduced with the use of membrane oxygenators instead of bubble oxygenators,[101,102] arterial line filtration,[9,102] and the nonuse of vacuum-assisted drainage of the shed blood during CPB.[100]

Beating Heart Surgery

While beating heart surgery provides more physiological conditions by eliminating some of the pathological features associated with CPB, it does not correlate with the expected benefits in neurological outcome. Arom et al[103] comparing 350 OPCAB to 3,171 CABG patients found a lower but not statistically significant difference for stroke (1.4% vs. 2.0%) and TIA (0.3% vs. 0.9%). However, the 2 groups were quite dissimilar. OPCAB patients had a significantly higher predicted risk (4.3±7.4% vs. 2.6%±4.6%, p< 0.001). In a comparable study, Iaco et al reported a lower rate of mortality and neurological complications for OPCAB versus CPB (1.9% vs. 3.8% and 0.4% vs. 1.7%, respectively).[104] To investigate the outcome of coronary revascularization with and without CPB in a high-risk category of patients, such as octogenarians, Ricci et al[105] performed a retrospective nonrandomized study in 269 patients, 97 without CPB and 172 with CPB. In the OPCAB group, there was a lower graft-patient ratio (1.8 vs. 3.3, p=ns), greater freedom from post-operative complications (85.6% vs.75%; p=0.004) and a higher percentage of reoperations (16.5% vs. 4.7%). The incidence of stroke was 0% in the off-pump cohort compared with 9.3% in the CPB cohort, which was highly significant (p=0.0005).

Murkin and colleagues investigated neurobehavioural patterns following OPCAB surgery.[106] A series of 35 OPCAB and 33 conventional patients were compared at 5 days and at 3 months post-operatively. While there were no differences in age between the 2 cohorts, the numbers of coronary anastomoses per patient (1.1 vs. 3.2), total operating room time (4.4h vs. 3.98h) and extubation time (5.9 h and 20h) were dissimilar. No patient suffered from a stroke in either group. Cognitive dysfunction was lower in the OPCAB group with an incidence of 66% vs. 90% (p=0.025) at 5 days, and of 5% vs. 50% respectively (p=0.001) at 3 months post-operatively. In a small, randomized study, Diegler et al demonstrated that embolic signals, defined as high intensive transient signals and used for continuous monitoring of middle cerebral artery blood flow by transcranial Doppler ultrasound, were significantly lower in the off-pump group (p=0.001).[107] In the OPCAB group, no patients suffered from a decrease in cognitive score, while such was the case in 90% of conventional CPB patients. In a similar investigation, Bowsles and colleagues noted that OPCAB reduced detectable emboli by 2 orders of magnitude on average, but without any improvement in neurological outcomes compared to standard CABG.[108] Beating heart surgery decreases embolic load by minimizing surgical manipulation of the aorta, cannulation/decannulation, and by eliminating micoemboli generated from the pump circuitry.[109] However, parts of them still remain due to removal of the side-biting clamp employed for the construction of proximal anastomoses. A few prospective, randomized studies have examined the neuroprotective role of beating heart surgery versus standard CABG but have yielded inconsistent results.[107,110,111] Recently, Zamvar from Edinburgh published their findings in a prospective, randomized trial on neurocognitive changes at 1 and 10 weeks after off-pump versus conventional surgery.[112] Sixty patients were included in their investigation. The results were favoured OBCAB surgery. At 1 week, OPCAB patients had only 27% impairment compared to 63% for CPB patients (p=0.004); at 10 weeks, this difference still persisted with 10% versus 40% impairment (p=0.017).

Role of the Off-Pump "No Touch" Technique

Manipulation of the severely atherosclerotic aorta is believed to be the leading cause of atheroembolisms during CPB, with a stroke rate of 25% among patients with mobile plaques of the aortic arch.[16] Ultrasonography of the ascending aorta detects aortic atherosclerotic disease and eventually changes the surgical strategy. Modifications of the operative technique include avoidance of ascending aortic cannulation, cannulation either of the distal transverse arch, the femoral or axillary artery, avoidance of aortic cross-clamping by hypothermic fibrillatory arrest, avoidance of side-bite clamping for proximal anastomoses with unique aortic clamping time, construction of proximal anastomoses on inflow sites other than the ascending aorta, and total arterial revascularization.[112-121] Alternatively, more aggressive approaches, such as aortic endarterectomy or graft replacement, could substantially reduces atheroembolism from the diseased aorta, but increases the mortality rate. Stern et al found that aortic endarterectomy under circulatory arrest in patients with severe atherosclerosis of the aorta was an independent predictor of peri-operative stroke.[121] Such results confirm that intra-operative aortic manipulation of the diseased aorta should be minimized as much as possible. Although the innominate or the subclavian artery could be used as an alternative inflow site for proximal anastomoses, it does not avoid the risk of stroke because of the prevalence of atheromatous disease in these vessels is high.[123]

The best strategy to prevent the risk of atheroembolism in patients with atherosclerosis of the aorta is to combine any manipulation of the aorta with the avoidance of CPB, such as the off-pump, "no touch" technique.[124] Avoiding CBP eliminates the need for aortic cannulation and decreases the risk of microembolization. To be optimal, however, proximal anastomosis has to be avoided by deploying pedicled arterial grafts (LITAs, RITAs, right gastroepiploic artery), or by connection of vascular conduits to a pedicled graft in a "Y" or "T" configuration.[125,126] Concerns have been expressed about the capacity of LITAs to provide adequate flow to the entirely ischemic heart, although there is accumulating evidence in the literature that they are capable of supplying more than one coronary territory by increasing flow reserves.[127]

We have reviewed our experience with the off-pump, "no touch" technique in 78 patients and compared them to standard off-pump surgery performed in 422 patients.[128] While both groups had similar demographics and preoperative risk factors, atherosclerosis of the aorta (0% vs. 13%, p=0.001), carotid disease (25% vs. 16%, p=0.05), and previous neurological accidents (15% vs. 8.6%, p=0.05) were more frequent in the "no-touch" patients. Incomplete revascularization was present in 10% of the no-touch group, and in 4.5% of the off-pump group. The mortality rate at 30 days was 1.7% and 1.6%, respectively, and no differences were found in low cardiac output syndrome, peri-operative MI, reexploration for bleeding, and mechanical ventilation time. In the "no-touch" patients, the incidence of delirium (9% vs. 18%, p=0.07), atrial fibrillation (18% vs. 29%, p=0.05) and stroke (0% vs. 1%, p=0.6) was in favor of OPCAB surgery along with shorter intensive care unit stay (50±38 hours vs. 74±110 hours, p=0.05). However, when we corrected for confounding variables, probably due to the small number of patients, the technique was found to have no direct benefit. We concluded that in the presence of severe atherosclerosis of the aorta, the off-pump "no-touch" technique combined the advantage of eliminating the deleterious effect of CPB with avoidance of aortic manipulation, resulting in a very low incidence of neurological complications. The technique was also discovered to be safe and secure. The long-term results of this technique remain to be reported.

Conclusion

Neurological complications are still a major concern in cardiac surgery. Many approaches have been developed to help prevent their devastating consequences. Off-pump surgery may not provide all the answers, but is one of those "in progress" strategies that could minimize them significantly.

References

1. Mora Mangano CT, Mangano DT. Perioperative stroke, encephalopathy and CNS dysfunction. J Intensive Care Med 1997; 12:148-60.
2. Mora CT, Murkin JM. The central nervous system: Responses to cardiopulmonary bypass. In: Mora CT, ed. Cardiopulmonary Bypass: Principles and Techniques of Extracorporeal Circulation. New York: Springer-Verlag, 1995:114-46.
3. Shaw PJ, Bates D, Cartlidge NEF et al. Early neurological complications of coronary artery bypass surgery. BMJ 1985; 291:1384-7.
4. Sotaniemi KA. Cerebral outcome after extracorporeal circulation: Comparison between prospective and retrospective evaluations. Arch Neurol 1983; 40:75-7.
5. Roach GW, Kanchuger M, Mangano CM et al. Adverse cerebral outcome after coronary bypass surgery. N Engl J Med 1996; 335:1857-63.
6. Wolman RL, Nussmeier NA, Aggarwal A et al. Cerebral injury after cardiac surgery: Identification of a group at extraordinary risk. Multicenter Study of Perioperative Ischemia Research Group (McSPI) and the Ischemia Research Education Foundation (IREF) Investigators. Stroke 1999; 30:514-22.
7. Mangano DT. Perioperative cardiac morbidity. Anesthesiology 1990; 72:153-84.
8. Clark RE, Brillman J, Davis DA et al. Microemboli during coronary artery bypass grafting: Genesis and affect on outcome. J Thorac Cardiovasc Surg 1995; 109:249-58.
9. Pugsley W, Klinger L, Paschalis C et al. The impact of microemboli during cardiopulmonary bypass on neuropsychological functioning. Stroke 1994; 25:1393-9.
10. Blauth CI, Cosgrove DM, Webb BW et al. Atheroembolism from the ascending aorta. J Thorac Cardiovasc Surg 1992; 103:453-62.
11. Barbut D, Hinton RB, Szatrowski TP et al. Cerebral emboli detected during bypass surgery are associated with clamp removal. Stroke 1994; 25:2389-402.
12. St-Amand MA, Murkin JM, Menkis AH et al. Aortic atherosclerotiques plaque identified by epiaortic scanning predicts cerebral embolic load in cardiac surgery (Abstract). Can J Anaesth 1999; 44:A7.
13. Davila-Roman VG, Barzilai B, Wareing TH et al. Intraoperative ultrasonographic evaluation of the ascending aorta in 100 consecutive patients undergoing cardiac surgery. Circulation 1991; 84(Suppl.):47-53.
14. Ricotta JJ, Faggioli GL, Castilone A et al. Risk factors for stroke after cardiac surgery: Buffalo Cardiac–Cerebral Study Group. J Vasc Surg 1995; 21:359-64.

15. Tuman KJ, McCarthy RJ, Najafi H et al. Differential effects of advanced age on neurologic and cardiac risks of coronary operations. J Thorac Cardiovasc Surg 1992; 104:1510-7.
16. Katz ES, Tunick P, Rusineck H et al. Protruding aortic atheromas predict stroke in elderly patients undergoing cardiopulmonary bypass; experience with intraoperative transesophageal echocardiography. J Am Coll Cardiol 1992; 20:70-7.
17. Wareing TH, Davila-Roman VG, Barzilai B et al. Management of the severely atherosclerotic ascending aorta during cardiac operations: A strategy for detection and treatment. J Thorac Cardiovasc Surg 1992; 103:453-62.
18. Murkin JM, Menkis A, Adams SJ et al. Epiaortic scanning significantly decreases intervention-related and total cerebral emboli. Heart Surg Forum 2003; 6(4):203-4.
19. Nakayama K, Yamamuro A, Ikuno Y et al. Evaluation of patients with cerebral infarction using tranesophageal echocardiography: Atherosclerotic changes in the thoracic aorta and the branches of the aortic arch. J Cardiol 1998; 32:21-30.
20. Borger M, Ivanov J, Weisel RD et al. Stroke during coronary bypass surgery: Principal role of cerebral macroemboli. Eur J Cardiothorac Surg 2001; 19(5):627-32.
21. Gardner TJ, Horneffer PJ, Manolio TA et al. Stroke following coronary artery bypass grafting: A ten–year study. Ann Thorac Surg 1985; 40:574-81.
22. Reed GL, Singer DE, Picard EH et al. Stroke following coronary artery bypass surgery. N Engl J Med 1988; 319:1246-50.
23. Breuer AC, Furlan AJ, Hanson MR et al. Central nervous system complications of coronary artery bypass graft surgery: Prospective analysis of 421 patients. Stroke 1983; 14:682-7.
24. Roper AH, Wechsler LR, Wilson LS. Carotid bruit and the risk of stroke in elective surgery. N Engl J Med 1982; 307:1388-90.
25. Horst M, Geissler HJ, Mehlhorn U et al. Simultaneous carotid and coronary artery surgery: Indications and perioperative outcome. J Thorac Cardiovasc Surg 1999; 47(5):328-32.
26. Das SK, Brown TD, Pepper J. Continuing controversy in the management of concomitant coronary and carotid disease: An overview. Int J Cardiol 2000; 74(1):47-65.
27. Fode NC, Sundt Jr TM, Robertson JT et al. Multicenter retrospective review of results and complications of carotid endarterectomy in 1981. Stroke 1986; 17:370-6.
28. North American Symptomatic Carotid endarterectomy Trial Collaborators. Beneficial effect of carotid endarterectomy in symptomatic patients with high-grade carotid stenosis. N Engl J Med 1991; 325:445-63.
29. Johnson LW, Dickstein RA, Fruehan CT et al. Prophylactic digitalization for coronary artery bypass surgery. Circulation 1976; 53:819-22.
30. Creswell LL, Scuessler RB, Rosenbloom M et al. Hazards of postoperative atrial arrhytmias. Ann Thorac Surg 1993; 56(3):405-9.
31. Leitch JW, Thomson D, Baird DK et al. The importance of age as predictor of atrial fibrillation and flutter after coronary artery bypass grafting. J Thorac Cardiovasc Surg 1990; 100:338-42.
32. Tyras DH, Stothert MD, Kaiser GC et al. Supraventricular tachyarrhytmias after myocardial revascularization: A randomized trial of prophylactic digitalization. J Thorac Cardiovasc Surg 1979; 77:310-4.
33. Roffman JA, Fieldman A. Digoxin and propanolol in the prophylaxis of supraventricular tachydysrhytmias after coronary artery bypass surgery. Ann Thorac Surg 1981; 31:496-501.
34. Nyström U, Edvardsson N, Berggren H et al. Oral sotalol reduces the incidence of atrial fibrillation after coronary artery bypass surgery. Thorac Cardiovasc Surg 1993; 41:34-7.
35. Roberts SA, Diaz C, Nolan PE et al. Effectiveness and costs of digoxin treatment for atrial fibrillation and flutter. Am J Cardiol 1993; 72:567-73.
36. Treggiari-Venzi MM, Weber JL, Perneger TV et al. Intravenous amiodarone or magnesium sulphate is not cost-beneficial prophylaxis for atrial fibrillation after coronary artery bypass surgery. Br J Anaesth 2000; 85(5):690-5.
37. Redle JD, Khurana S, Marzan R et al. Prophylactic oral amiodarone compared with placebo for prevention of atrial fibrillation after coronary artery bypass surgery. Am Heart J 1999; 138(1Pt 1):144-50.
38. Stroke prevention in atrial fibrillation study group investigators. Preliminary report of the Stroke Prevention in Atrial Fibrillation Study. N Engl J Med 1990; 322:863-8.

39. The boston area anticoagulation trial for atrial fibrillation investigators. The effect of low-dose warfarin on the risk of stroke in patients with nonrheumatic atrial fibrillation. N Engl J Med 1990; 323:1505-1511.

40. Opie JC. Cardiac surgery and acute neurological injury. In: Willner A, ed. Cerebral Damage Before and After Cardiac Surgery. Dordrecht: Kluwer Academic Publishers, 1993:15-36.

41. Trzepacz PT. The neuropathogenesis of delirium. A need to focus our research. Psychosomatics 1994; 35:374-91.

42. van der Marst RC, van der Broek WW, Fekkes D et al. Incidence of and preoperative predictors for delirium after cardiac surgery. J Psychosom Res 1999; 46(5):479-83.

43. Franco K, Litaker D, Locala J et al. The cost of delirium in the surgical patient. Psychosomatics 2001; 42:68-73.

44. Rockwood K. The occurrence and duration of symptoms in elderly patients with delirium. J Gerontol 1993; 48:M162-6.

45. Francis J, Martin D, Kapoor WN. A prospective study of delirium in hospitalized elderly. JAMA 1994; 271:134-9.

46. Levkoff SE, Evans DA, Lliptzin B et al. Delirium. The occurrence and persistence of symptoms among the elderly hospitalized patients. Arch Intern Med 1992; 152:334-40.

47. Francis J, Martin D, Kapoor WN. A prospective study of delirium in hospitalized elderly. JAMA 1990; 263:1097-101.

48. Francis J, Kapoor WN. Prognosis after hospital discharge of older medical patents with delirium. J Am Geriatr Soc 1992; 40:601-6.

49. Koponem H, Stenback U, Mattila E et al. Delirium among elderly persons admitted to a psychiatric hospital: Clinical course during acute stage and one-year follow-up. Acta Psychiatr Scand 1989; 79:579-85.

50. Rolfson DB, McElhaney JE, Rockwood K et al. Incidence and risk factors for delirium and other adverse outcomes in older adults after coronary artery bypass graft surgery. Can J Cardiol 1999; 15(7):771-6.

51. Brown WR, Moody DM, Challa VR et al. Longer duration of cardiopulmonary bypass is associated with greater numbers of cerebral microemboli. Stroke 2000; 31:707-13.

52. Gokgoz L, Gunaydin S, Sinci V et al. Psychiatric complications of cardiac surgery postoperative delirium syndrome. Scand Cardiovasc J 1997; 31(4):217-22.

53. Breuer AC, Furlan AJ, Hanson MR et al. Central nervous system complications of coronary artery bypass graft surgery: Prospective analysis of 421 patients. Stroke 1983; 14:682-7.

54. Mangano DT. Cardiovascular morbidity and CABG surgery, a perspective: Epidemiology, costs, and potential therapeutics solutions. J Cardiovasc Surg 1995; 10(Suppl):366-8.

55. Smith PLC, Newman SP, Ell PJ et al. Cerebral consequences of cardiopulmonary bypass. Lancet 1986; 1:823-5.

56. Newman S. The incidence and nature of neuropsychological morbidity following cardiac surgery. Perfusion 1989; 4:93-100.

57. Savageau JA, Stanton BA, Jenkins CD et al. Neuropsychological dysfunction following elective cardiac operation. Early assessment. I. J Thorac Cardiovasc Surg 1982; 84:585-94.

58. Shaw PJ, Bates D, Cartlidge NEF et al. Neurologic and neuropsychological morbidity following major surgery: Comparison of coronary artery bypass and peripheral vascular surgery. Stroke 1987; 18:700-7.

59. Savageau JA, Stanton BA, Jenkis CD et al. Neuropsychological dysfunction following elective cardiac operations. A six month reassessment. II. J Thorac Cardiovasc Surg 1982; 84:595-600.

60. Shaw PJ, Bates D, Cartlidge NEF et al. Long-term intellectual dysfunction following coronary artery bypass graft surgery: A six-month follow-up study. Q J Med 1987; 239:259-68.

61. Newman M, Kirchner JL, Phillips-Bute B et al. Longitudinal assessment of neurocognitive function after coronary artery bypass surgery. N Engl J Med 2001; 344(6):395-401.

62. Selnes OA, Royall RM, Grega MA et al. Cognitive changes 5 years after coronary artery bypass grafting. Arch Neurol 2001; 58:598-604.

63. Howard R, Trend P, Ross Russell RW. Clinical features of ischemia in cerebral arterial border zones after periods of reduced cerebral blood flow. Arch Neurol 1987; 44:934-40.

64. Degirmenci B, Durak H, Hazan E et al. The effect of coronary artery bypass surgery on brain perfusion. J Nucl Med 1998; 39:587-91.
65. Murkin JM, Martzke JS, Buchan AM et al. A randomized study on the influence of perfusion technique and pH management strategy in 316 patients undergoing coronary artery bypass surgery: Mortality and cardiovascular morbidity. J Thorac Cardiovasc Surg 1995; 110:340-8.
66. Gilman S. Cerebral disorders after open-heart operations. N Engl J Med 1965; 272:489-98.
67. Selnes OA, Goldsborough MA, Borowicz LM et al. Neurobehavioral sequelae of cardiopulmonary bypass. Lancet 1999; 353(9164):1601-6.
68. Newman MF, Croughwell ND, Blumenthal JA et al. Predictors of cognitive decline after cardiac operation. Ann Thorac Surg 1995; 59:1326-30.
69. Moody DM, Bell MA, Johnston WE et al. Brain microemboli during cardiac surgery or aortography. Ann Neurol 1990; 28:477-86.
70. Brooker RF, Brown WR, Moody DM et al. Cardiotomy suction: A major source of brain lipid emboli during cardiopulmonary bypass. Ann Thorac Surg 1998; 65:1651-5.
71. Kincaid EH, Jones TJ, Stump DA et al. Processing scavenged blood with a cell saver reduces cerebral lipid microembolization. Ann Thorac Surg Oct 2000; 70(4):1296-300.
72. Stump DA, Rogers AT, Hammon JW et al. Cerebral emboli and cognitive outcome after cardiac surgery. J Cardiovasc Anesth 1996; 1:113-9.
73. Kramer DC, Stanley TE, Sanderson I et al. Failure to demonstrate relationship between mean arterial pressure during cardiopulmonary bypass and postoperative dysfunction (Abstract). Anesthesiology 1994; 81:A156.
74. Shell R, Croughwell N, White W et al. The effect of time and rewarming from moderate hypothermic cardiopulmonary bypass on cerebral blood flow. Anesthesiology 1992; 77:A55 Abstract.
75. Schell RM, Stanley T, Croughwell N et al. Temperature during cardiopulmonary bypass and neuropsychologic outcome. Anesthesiology 1992; 77:A119 Abstract.
76. Brown WR, Moody DM, Tytell M et al. Microembolic brain injuries from cardiac surgery: Are they seeds of future Alzheimer's disease? Ann NY Acad Sci 1997; 826:386-9.
77. Tardiff BE, Newman MF, Saunders AM et al. Preliminary report of a genetic basis for cognitive decline after cardiac operations. Ann Thorac Surg 1997; 90; 715-20.
78. Frasure-Smith N, Lesperance F, Talajiic M. Depression following myocardial infarction: Impact on 6-month survival. JAMA 1993; 270:1819-25.
79. Langeluddecke PM, Fulcher G, Baird D et al. A prospective evaluation of the psychosocial effects of coronary artery bypass surgery. J Psychosom Res 1989; 33:37-45.
80. McKhann GM, Borowicz LM, Goldsborough MA et al. Depression and cognitive decline after coronary artery bypass. Lancet 1997; 349:1282-84.
81. JK Kirklin, Westaby S, Blackstone EH et al. Complement and damaging effects of cardiopulmonary bypass. J Thorac Cardiovasc Surg 1983; 86:845-57.
82. Kirklin JK. Prospects for understanding and eliminating the deleterious effects of cardiopulmonary bypass (Editorial). Ann Thorac Surg 1991; 51:529-31.
83. Butler J, Rocker GM, Westaby S. Inflammatory response to cardiopulmonary bypass. Ann Thorac Surg 1993; 55:552-9.
84. Harris DNF, Bailey SM, Smith PLC et al. Brain swelling in the first hour after coronary artery bypass surgery. Lancet 1993; 342:586-7.
85. The warm heart investigators. Randomised trial of normothermic versus hypothermic coronary bypass surgery. Lancet 1994; 343(8897):559-63.
86. Martin TD, Craver JM, Gott JP et al. Prospective, randomised trial of retrograde warm blood cardioplegia: Myocardial benefit and neurologic threat. Ann Thorac Surg 1994; 57:298-304.
87. Croughwell ND, Newman MF, Blumenthal JA et al. Jugular bulb saturation and cognitive dysfunction after cardiopulmonary bypass. Ann Thorac Surg 1994; 58:1702-8.
88. Rahn H, Reeves RB, Howell BJ. Hydrogen ion regulation, temperature, and evolution. Am Rev Resp Dis 1975; 122:165-72.
89. Swan H. The importance of acid-base management for cardiac and cerebral preservation during open-heart operations. Surg Gynecol Obstet 1984; 158:391-414.
90. Anstadt MP, Tedder M, Hedge SS et al. Pulsatile versus nonpulsatile reperfusion improves cerebral blood flow after cardiac arrest. Ann Thorac Surg 1993; 56:453-61.

91. Prough DS, Stump DA, Roy RC et al. Response to cerebral blood flow to changes in carbon dioxide tension during hypothermic cardiopulmonary bypass. Anesthesiology 1986; 64:576-81.

92. Murkin JM, Farrar JK, Tweed WA et al. Cerebral autoregulation and flow/metabolism coupling during cardiopulmonary bypass: The influence of PaCO$_2$. Anesth Analg 1987; 66:825-32.

93. Stephan H, Weyland A, Kazmaier S et al. Acid-base management during hypothermic cardiopulmonary bypass does not affect cerebral metabolism but does affect blood flow and neurological outcome. Br J Anaesth 1992; 69:51-7.

94. Tufo HM,Ostfeld AM, Shekelle R. Central nervous system dysfunction following open heart surgery. JAMA 1970; 212:133-40.

95. Stockard JJ, Bickford RG, Schauble JF. Pressuredependent cerebral ischemia during cardiopulmonary bypass. Neurology 1973; 23:521-9.

96. Kolkka R, Hilberman M. Neurologic dysfunction following cardiac operation with low-flow, low-pressure cardiopulmonary bypass. J Thorac Cardiovas Surg 1980; 79:432-7.

97. Slogoff S, Girgis KZ, Keats AS. Etiologic factors in neuropsychologic complications associated with cardiopulmonary bypass. Anesth Analg 1982; 61:903-11.

98. Bashein G, Bledsoe SE, Townes BD et al. Tools for assessing central nervous system injury in the cardiac surgical patient. In: Hilberman M, ed Martinus Nijhoff, Boston: 1988:109-36.

99. Padayachee TS, Parsons S, Theobold R et al. The detection of microemboli in the middle cerebral artery during cardiopulmonary bypass: A transcranial Doppler ultrasound investigation using membrane and bubble oxygenators. Ann Thorac Surg 1987; 44:298-302.

100. Willcox T, Mitchell SJ, Gorman DF. Venous air in the bypass circuit: A source of arterial line microemboli exacerbated by vacuum-assisted drainage. Ann Thorac Surg 1999; 68:1285-9.

101. Blauth C, Smith PL, Arnold JV et al. Influence of oxygenator type on the incidence and extent of microembolic retinal ischemia during cardiopulmonary bypass. Assessment by digital image analysis. J Thorac Cardiovasc Surg 1990; 99:61-9.

102. Padayachee TS, Parsons S, Theobold R et al. The effect of arterial filtration on reduction of gaseous microemboli in the middle cerebral artery during cardiopulmonary bypass. Ann Thorac Surg 1988; 84(45):647-9.

103. Arom KV, Flavin TF, Every RW e al. Safety and efficacy of off-pump coronary artery bypass grafting. Ann Thorac Surg 2000; 69:704-10.

104. Iaco AL, Contini M, Teodori G et al. Off or on bypass: What is the safety threshold? Ann Thorac Surg 1999; 68:1486-90.

105. Ricci M, Karamanoukian H, Abreham R et al. Stroke in octogenarians undergoing coronary artery surgery with and without cardiopulmonary bypass. Ann Thorac Surg 2000; 69:1471-5.

106. Murkin JM, Boyd WD, Ganaphathy S et al. Postoperative cognitive dysfunction is significantly less after coronary artery revascularization without cardiopulmonary bypass. Ann Thorac Surg 1999; 68:1469 Abstract.

107. Diegeler A, Hirsch R, Schneider F et al. Neuromonitoring and neurocognitive outcome in off-pump versus conventional coronary bypass operation. Ann Thorac Surg 2000; 69:1162-6.

108. Bowsles BJ, Lee JD, Dang CR et al. Coronary artery bypass performed without the use of cardiopulmonary bypass is associated with reduced cerebral microemboli and improved clinical results. Chest 2001; 119:25-30.

109. Watters MPR, Cohen AM, Monk CR et al. Reduced cerebral embolic signals in beating heart coronary surgery detected by transcranial doppler ultrasound. Br J Anaesth 2000; 84:629-31.

110. Lloyd CT, Ascione R, Underwood MJ et al. Serum S-protein release and neuropsychologic outcome during coronary revascularization on the beating heart: A prospective randomized study. J Thorac Cardiovasc Surg 2000; 119:148-54.

111. Van Djick D, Jansen EW, Hijman R et al. The octopus study group. Cognitive outcome after off-pump and on-pump coronary artery bypass graft surgery: A randomized trial. JAMA 2002; 287:1448-50.

112. Zamvar V, Williams D, Hall J et al. Assessment of neurocognitive impairment after off-pump and on-pump techniques for coronary artery bypass graft surgery: Prospective randomized controlled trial. BMJ 2002; 325:1268-73.

113. Malheiros SMF, Massaro AR, Cavalho AC et al. Transesophageal echocardiography and transcranial Doppler monitoring in coronary surgery with and without cardiopulmonary bypass: Preliminary results. Cerebrovasc Dis 1999; 9:358-60.

114. Leyh RG, Bartels C, Notzold A et al. Management of porcelain aorta during coronary artery bypass grafting. Ann Thorac Surg 1999; 67:986-8.

115. Baribeau RY, Westbrook BM, Charlesworth DC et al. Arterial inflow via axillary artery graft for the severely atheromatous aorta. Ann Thorac Surg 1998; 66:33-7.

116. Sabix JF, Lytle BW, McCarthy PM et al. Axillary artery: An alternative site of arterial cannulation for patients with extensive aortic and peripheral vascular disease. J Thorac Cardiovasc Surg 1995; 109:885-91.

117. Salerno TA. Single aortic cross-clamping for distal and proximal anastomoses in coronary surgery: An alternative to conventional techniques. Ann Thorac Surg 1982; 33:518-20.

118. Aranki SF, Sullivan TE, Cohn L. The effect of single aortic cross-clamp technique on cardiac and cerebral complications during coronary bypass surgery. J Cardiovasc Surg 1995; 10:498-502.

119. Weinstein G, Killen DA. Innominate artery coronary artery bypass graft in a patient with calcific aortitis. J Thorac Cardiovasc Surg 1980; 79:312-3.

120. Calafiore AM, DiGiammarco G. Complete revascularization with three or more arterial conduits. Sem Thorac Cardiovasc Surg 1996; 8:15-23.

121. Culliford AT, Colvin SB, Rohrer K et al. The atherosclerotic ascending aorta and trasverse arch: A new technique to prevent cerebral embolization during bypass. Experience with 13 patients. Ann Thorac Surg 1986; 41:27-35.

122. Stern A, Tunick PA, Culliford AT et al. Protruding aortic arch atheromas: Risk of stroke during heart surgery with and without aortic arch endarterectomy. Am Heart J 1999; 138:746-52.

123. Tobler HG, Edwards JE. Frequency and location of atherosclerotic plaques in the ascending aorta. J Thorac Cardiovasc Surg 1988; 96:304-6.

124. Ricci M, Karamanoukian HL, D'Ancona G et al. Preventing neurologic complications in coronary artery surgery: The "off-pump, no touch" technique. Am Heart J 1999; 138:746-52.

125. Tector AJ, Amundsen S, Shmahl TM et al. Total revascularization with T grafts. Ann Thorac Surg 1994; 57:33-8.

126. Royse AG, Royse CF, Raman JS. Exclusive Y graft operation for multivessel coronary revascularization. Ann Thorac Surg 1999; 68:1612-8.

127. Wendler O, Henne B, Markwirth T et al. T grafts with the right internal thoracic artery to the left internal thoracic artery versus the left internal thoracic artery and radial artery: Flow dynamics in the left thoracic artery stem. J Thorac Cardiovasc Surg 1999; 118:841-8.

128. Leacche M, Carrier M, Bouchard D et al. Improving neurological outcome in off-pump surgery: The "no-touch" technique. Heart Surg Forum 2003; 6(3):169-75.

CHAPTER 11

OPCAB Surgery in High Risk-Patients

Jehangir J. Appoo and Raymond Cartier

The advent of cardiopulmonary bypass (CPB) circulation in the past half century has revolutionized the field of cardiac surgery. Although CPB has been associated with very low morbidity, its side-effects can be detrimental. The purported advantages of "off-pump" coronary surgery are potentially many. Namely these are reduced mortality and morbidity, including a decreased incidence of central nervous system events, myocardial reperfusion injury, bleeding, arrhythmias, respiratory insufficiency, renal insufficiency as well as shorter length of hospital stay and lower cost. The effects of CPB apparatus are not completely understood, but the inflammatory cascade stimulated by CPB may be poorly tolerated by the myocardium. For instance, patients with impaired left ventricular function undergoing conventional coronary artery bypass grafting (CABG) with CPB are still at increased risk of peri-operative mortality compared to subject with normal left ventricular function.[1-3] Thus, logically, patients at high risk from CPB may be expected to show improved outcomes after OPCAB. Traditional "high-risk" patients for conventional CABG on CPB have included those with decreased left ventricular function (LVEF < 35%), redo surgery, increased age, left main stem stenosis, recent myocardial infarction (MI), pulmonary hypertension, cerebrovascular disease, and chronic renal failure. Classically, high-risk patients are excluded from randomized trial, and have not been included in recent prospective, randomized trials comparing on- and off-pump bypass.[4-7] Paradoxically, they represent a constantly-growing segment of the surgical population. We review these risk factors and how they can interact with OPCAB surgery.

OPCAB in Patients with Impaired LVEF

Several studies have found reduced LVEF to be a significant risk factor for peri-operative mortality after conventional CABG.[1-3] Arom et al[8] compared a contemporary group of 177 patients with LVEF ≤ 30% undergoing CABG. Forty-five patients underwent OPCAB, and the remaining 132, conventional coronary artery bypass surgery (CCAB). Operative mortality in the OPCAB group was 4.4% vs. 7.7% in the CCAS group. However, this did not achieve statistical significance because of the small patient population. Blood loss and peak CK-MB were lower in the OPCAB group. They concluded that multivessel OPCAB in patients with left ventricular dysfunction was appropriate and applicable. Meharwal and Trehan[9] compared 355 patients with multivessel coronary artery disease and LVEF ≤ 30% undergoing OPCAB to a contemporary and similar group of 959 patients undergoing CCAB. Peri-operative mortality was 3.9% and 6.0% in the off-pump and on-pump groups respectively, but the difference did not reach statistical significance (p=0.18). Post-operative morbidity was reduced in the off-pump group with statistically significant decline in the incidence of atrial fibrillation and prolonged ventilation. Intensive care unit (ICU) and hospital length of stay were also decreased in the off-pump group. Both groups of patients were followed up for 4 yrs post-op (mean 18 ± 12

Table 1. Risk factors for postoperative complication "cumulative score index" on patients with a LVEF ≤ 0.35% operated either under cardiopulmonary bypass (CPB) or off-pump (OPCAB).

	P Value	O.R.
Female sex:	0.001	1.54
Age:	0.003	1.05
IABP (preoperative):	0.001	8.57
Peripheral vascular disease:	0.001	2.39
Surgical technique[#]:	0.022	1.4

[#] Reference value: OPCAP = 0; IABP= Intra-aortic balloon pump device

months), and there was no difference in the incidence of late death, recurrent angina, nonfatal MI or reintervention. Thus, they concluded that post-operative morbidity was lower in the OPCAB group with no compromise of long-term outcome. In early experience with OPCAB, Moshkovitz and associates[10] reported their results from 1991-94 in 75 patients with LVEF ≤ 35.[10] The mean number of grafts in this early study was 1.9 grafts/patient with only 23% receiving a graft to the circumflex territory. The patient population included a high percentage with congestive heart failure (CHF), evolving MI or cardiogenic shock. Peri-operative mortality was 2.7%, and 1 and 4 year actuarial survival was 96% and 73%, respectively.

In our series of 750 consecutive and systematic patients who had undergone OPCAB surgery, 77 had a preoperative LVEF ≤ 35%. Although operative mortality was lower than that one observed in 142 contemporary patients with similar conditions (3.8 vs 6.5%, p = 0.45), statistical significance was not reached. Logistic regression analysis performed on the entire cohort revealed that the independent risk factors were female sex, left main disease, chronic renal insufficiency, and preoperative use of intra-aortic balloon pump (IABP). The surgical technique was not identified as an independent risk factor (P = 0.27, odds ratio = 2.8). However, when post-operative complications (death, peri-operative MI, post-operative IABP, hospital stay > 7 days) were assessed by complication cumulative index, the outcome was in favour of OPCAB patients (Table 1).

OPCAB in Patients Undergoing Reop Surgery

Reoperative CABG is definitely considered high-risk surgery. These patients are generally older and the "resternotomy" is generally seen as a significant hazard that could jeopardize right ventricular myocardium integrity as well as previous patent bypasses.[11] Mortality is generally higher. Although avoiding extracorporeal circulation could help to decrease morbidity, it cannot prevent myocardial injury during chest reopening. In 1993, Fanning and colleagues[12] reported on 59 patients undergoing coronary reop without CPB, representing 10% of all reop surgeries. All the patients had single or double vessel disease, and were approached either through a middle sternotomy or the left chest, depending of the location of the target vessels. Peri-operative outcome was excellent, with no deaths, and in 20 patients who required a second coronary angiogram for recurrent chest pain, 18 of 20 grafts were patent. At 42 months 70% of the patients were in functional class I or II. These results are outstanding considering that these bypass graftings were carried out without coronary stabilization.

Trehan and associates[13] published their results on 50 patients undergoing redo OPCAB with low peri-operative morbidity and mortality, and satisfactory graft patency.[13] However, the majority of the patients had single and double vessel disease (48 patients). Operative mortality

Table 2. Risk factors for postoperative complication "cumulative score index" among patients who underwent reoperative coronary surgery operated either under cardiopulmonary bypass (CPB) or off-pump (OPCAB).

	P Value	O.R.
Age:	0.047	1.05
LVEG:	0.072	0.07
IABP (preoperative):	0.002	13.45
Peripheral vascular disease:	0.014	3.04
Surgical technique*:	0.042	2.34

*Reference value: OPCAB = 0; IABP= Intra-aortic balloon pump device

was 4%, and peri-operative MI, 2%. Stamou and colleagues[14] compared a series of 41 on-pump cases to 91 off-pump cases.[14] All had reoperative coronary surgery for single vessel disease. Hospital mortality was much higher in the on-pump group (10% versus 1%, p=0.003), which for single vessel disease was far beyond expectation. Yau and colleague[15] from the Toronto General Hospital have reviewed their experience with on pump coronary reoperation from 1992 to 1997. Out of 20,614 patients operated on during that period, 6.0% were reoperated. Compared to patients undergoing primary operation, those who underwent reoperation had a higher risk of peri-operative MI (3.7 vs 7.4%), low output syndrome (9 vs 24%), and death (2.4% vs 6.8%). Major predictors of mortality were aging, preoperative symptoms, earlier year of operation, and left ventricular dysfunction. With time, the surgical risk profile tended to increase, but morbidity and mortality remained stable.

At the Montreal Heart Institute, 61 (8%) and 99 patients (7.8%) from our cohorts of 750 systematic OPCAB and 1444 CPB patients respectively, underwent coronary reoperative surgery between 1997 and 1998. Although the preoperative data were comparable for both groups, there was a trend toward higher operative mortality in CPB patients (6.1 vs 1.6%, p = 0.34) and peri-operative MI (12.1 vs 8.2%, p = 0.65). Nevertheless, after normalization for preoperative risk factors, the cumulative complications index (operative mortality, peri-operative MI, hospital stay > 7 days, and post-operative insertion of an IABP) favoured OPCAB patients (Table 2).

Although the clinical experience with reoperative CABG surgery is still modest and the data are scattered in a few disparate studies, a common trend is emerging towards a better outcome with OPCAB surgery. This, if confirmed in the future, will have to be taken seriously by cardiac surgeons confronting reoperative surgery.

OPCAB in the Elderly

Advanced age is generally associated with increased comorbidity as well as post-operative morbidity and mortality.[16,17] The elderly are also the fastest growing segment of our surgical population. The patient population over age 70 years has grown from 15 to 25% between 1993 and 1997 according to the Ontario Cardiac Care Network.[18] The elderly are natural target for less invasive techniques, such as off pump surgery to decrease morbidity and mortality; Stamou and colleagues[19] analyzed their results in 71 octogenarians undergoing OPCAB, and found an operative mortality rate of 6%. According to their literature review, this compares favourably with published mortality rates of 6-13% in octogenarians undergoing on-pump CABG. They also showed, in their OPCAB population, that atrial fibrillation, pneumonia, inotropic support, length of hospital stay, and hospital death were more common than in

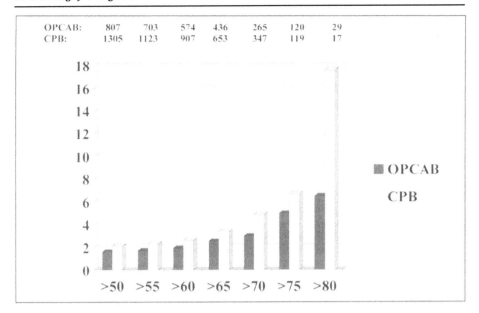

| OPCAB: | 807 | 703 | 574 | 436 | 265 | 120 | 29 |
| CPB: | 1305 | 1123 | 907 | 653 | 347 | 119 | 17 |

Figure 1. Operative mortality according to patient age and surgical technique.

younger (< 80-years old) OPCAB patients. Demers and Cartier[20] reported a lower incidence of atrial fibrillation and blood transfusion in a group of 98 patients over the age of 70 years undergoing OPCAB, compared to a noncontemporary group of 497 CPB patients. Operative mortality (3.1% OPCAB, 3.6% CPB) and the rate of neurological events (1% OPCAB, 3.2% CPB, p=0.4) were comparable for the 2 groups.

Koutlas et al[21] compared 53 patients aged 75 years or over undergoing OPCAB with 220 patients who underwent CCAB during the same time period. Mean patient age for both groups was 79 ± 0.5 years.[21] Mortality in the OPCAB group was 0% compared to 7.6% in CCAB (p < .05). Length of stay and transfusion requirements were decreased in the OPCAB group. There was no difference in the incidence of neurological complications, peri-operative MI, atrial fibrillation, post-operative hemorrhage, and renal failure. However, the mean number of grafts was much lower in the OPCAB (1.5 ±.1) vs. CCAB (2.5 ±.01) indicating either incomplete revascularization or decreased incidence of multivessel disease in this group. We evaluated operative mortality according to the age range in our cohort of 900 systematic OPCAB patients (operated on between September 1996 and May 2003) and compared it to our cohort of conventional cases. The results are shown in Figure 1. Although not significant, a definite trend toward lower mortality was observed in the OPCAB cohort. These results reflect the common trend in the literature and suggest that elderly patients might be those who could benefit the most from OPCAB surgery.[22,23]

Risk Adjusted Comparison Between OPCAB and CCAB

In a recent paper by Sabik et al[24] from the Cleveland Clinic, a sophisticated technique of boot strapping and propensity score matching was employed to match 406 OPCAB patients with 406 CCAB patients. In this very closely matched patient population, hospital mortality was similar (0.5% OPCAB vs. 1% CCAB). Overall morbidity was reduced in the OPCAB group with statistically significant decreases in the incidence of encephalopathy, sternal wound infection, red blood cell transfusion, and renal failure requiring dialysis. However, the mean number of grafts in the OPCAB cohort was lower (2.8 ± 1.0 vs. 3.5 ± 1.1), introducing some uncertainty about long-term results.

Magee and colleagues[25] showed that despite increased predicted risk among 1,983 patients undergoing OPCAB at their centre, operative mortality was significantly lower, 1.8% in OPCAB vs. 3.5% in CCAB. OPCAB patients were also found to have a decreased incidence of blood transfusion, prolonged ventilation, reoperation for bleeding, and reduced length of hospital stay. In a retrospective, multi-centre Society of Thoracic Surgeons database review, Cleveland compared almost 12,000 patients undergoing OPCAB surgery to over 105,00 CABG only procedures done on CPB.[26] Overall risk-adjusted mortality was found to be lower in the OPCAB group (2.3 vs. 2.9%, p< 0.0001). OPCAB decreased the risk-adjusted major complication rate from 14.2% to 10.6% (p< 0.0001). A similar study compared OPCAB in 9 Veterans Administration centres to CCAB in 34 centres using Deparment of Veterans Affairs Continuous Improvement in Cardiac Surgery Program records.[27] Again, patients treated off-pump had lower risk-adjusted mortality and morbidity (p< 0.05).

In a comparison of over 2,200 OPCAB patients at 8 UK centres with over 5,000 patients undergoing CCAB at the same institutes, the OPCAB group had a significantly reduced rate of operative mortality and morbidity.[28] Bitner and Savitt[29] attempted to look at this group of patients by reviewing 57 cases with markedly increased Parsonnet scores undergoing OPCAB in a 2-year period. The 30-day mortality rate was 3.5% compared to a predicted mortality of 10-15% in this patient population. However, no direct comparison was made with a group of patients undergoing CCAB at their institution. These studies have demonstrated that OPCAB is feasible in traditionally high-risk patients with a decreased incidence of major morbidity and a tendency towards improved operative mortality. No multicentre trial has yet directly compared patients at high risk for CABG undergoing OPCAB vs. CCAB. In recent randomized study conducted at the Montreal Heart Institute on high-risk patients, significant benefit was found to favour patients operated off-pump.[30] A total of 65 patients were randomized to on- or off-pump surgery. Inclusion criteria were at least 3 of the following: age greater than 65 years, high blood pressure, diabetes, serum creatinine >133 mmol/L, LVEF < 45%, chronic pulmonary obstructive disease, unstable angina, congestive heart failure, redo CA BG, anemia, and carotid atherosclerosis. Although there was a trend to lower operative mortality in the OPCAB cohort (0% vs 7%; p=0.1), the combined morbidity end-point (operative death, neurological injury, respiratory failure, operative MI) was significantly higher in CCAB patients (p=0.02). These results are quite encouraging, and should convince surgeons to be more aggressive in the use of off-pump surgery in this subset of patients.

Intra-Aortic Balloon Pump in High-Risk OPCAB Patients

Uncontrolled ischemia with or without hemodynamic instability remains the major obstacle to off-pump coronary artery revascularization. IABP is a very efficient means of normalizing myocardial ischemia and hemodynamic instability, and has served for more that 2 decades in conventional coronary surgery as well as in ICU cardiology.[31] Its has been very useful for unstable patient undergoing off-pump surgery.[32]

Kim et al[33] reviewed 142 consecutive OPCAB cases, 57 of whom received an IABP peror intra-op and 85 patients who did not receive an IABP. The mean duration of IABP support was 6.7 ± 9.5 hours. IABP helped to achieve an equivalent number of distal anastomosis in each group (3.4 vs. 3.5), despite the fact that the IABP group had a higher percentage of left main stenosis, unstable and intractable angina, left ventricular dysfunction, and urgent cases. There was no significant difference in mortality or major post-operative morbidity in the 2 groups. The only IABP-related complication occurred in a patient with calcified femoral arteries. In this case, limb ischemia was treated successfully with embolectomy. The authors concluded

Table 3. Risk factors for postoperative complication "cumulative score index" among OPCAB patients who IABPimplantation prior to surgery

	P value	O.R.
Age:	0.000	1.04
Sex:	0.038	1.53
Peripheral vascular disease:	0.025	1.64
LVEG:	0.007	0.16
Renal Insufficiency:	0.014	3.26
IABP (preoperative):	0.003	2.63

IABP= Intra-aortic balloon pump device

that high-risk patients can undergo complete revascularization via OPCAB with the same morbidity and mortality as low-risk patients.

Craver and Murrah[34] published their results on 16 patients undergoing high-risk OPCAB with IABP. There were no complications with IABP use, and they noted a decreased need for inotropic support during displacement of the heart. In a small series of high-risk patients (defined by LVEF < 35% and/or significant left main) Babatasi et al[35] explored the value of IABP prior to surgery. Out of 23 OPCAB procedures meeting these criteria, 70% benefited from presurgical IABP insertion, whereas only 46% of 15 CPB patients had the device put in prior to the procedure. No device-related complications were reported. Operative mortality was significantly lower among OPCAB patients with a trend toward decreased morbidity compared to patients operated conventionally.

In our cohort of 750 consecutive OPCAB patients at the Montreal Heart Institute, 65 had an IABP inserted prior to surgery for either unstable hemodynamics or uncontrolled ischemia. Subjects with an IABP had a higher Parsonnet risk stratification score (19 + 9 vs 11+8; p< 0.001), and were also older (66 ± 9 vs 64 ± 10; p = ns). There was a trend towards higher operative mortality (1.8 vs 1.2%; p = 0.87), and the complication cumulative score index (stay > 7 days, operative mortality, peri-operative MI, use of inotropes > 48 hours) was higher in the IABP group (61 vs 33%, p< 0.001). By stepwise logistic regression analysis, preoperative IABP was not identified as a significant risk factor for mortality, but was a significant factor for a positive complication cumulative score index (Table 3).

The above-cited studies confirm that IABP can be used safely to facilitate OPCAB in high-risk patients.

OPCAB Surgery in Acute Coronary Artery Syndrome (ACS)

The idea of performing direct revascularization on the acutely ischemic myocardium is not new. In the early 1960s, Kolesov[36] had already experimented successfully with the technique. Benetti and colleagues[37] reported in 1996 on a series of 32 patients operated while in acute myocardial infarction (AMI). They had no deaths and no clinical episodes of low cardiac output. However, the timing of the operation was not clearly specified. Wos et al[38] published their findings on a larger series of primary OPCAB patients presenting with AMI. They revascularized 165 consecutive "troponine-positive" AMI patients with a mortality rate of 4.2%. Among those in whom surgery had to be performed in the first 72 hours following the AMI, mortality rosed to 10.9%. Independent risk factors were female gender, early revascularization (< 72 hours), left ventricular aneurysm, peri-operative IABP, and right ventricular failure. Although the authors claimed the OPCAB approach was valuable, they underlined the significant mortality and morbidity associated with it, especially when the procedure was undertaken within 72 hours, which suggests that mortality increases with the severity and urgency of OPCAB. Ochi and associates[39] from Japan reported their experience on a similar topic. They compared 25 off-pump to 47 on-pump patients revascularized as an emergency for ACS. OPCAB patients received fewer grafts (2.6 vs. 3.8, p<0.0001). Mortality was comparable (on-pump: 8.5% vs. off-pump: 12%). They included multi-vessel patients or those with critical left main stenosis. According to multivariate analysis, the choice of procedure did not correlate with operative mortality. The authors concluded that OPCAB could be safe in selected ACS patients necessitating emergent CABG.

At this point, the place of OPCAB surgery in ACS is still not clearly defined. Previous studies did not put in perspective the potential unstable hemodynamic aspect that quite frequently affects these patients and which could be a limiting aspect of the procedure.

Conclusion

The results from various centres across the world have confirmed the short-term safety and efficacy of OPCAB in high-risk patients. Many of these studies have shown improvements in operative mortality or trends towards improved outcomes with OPCAB when compared to conventional CABG on CPB. However, due to the nature of this category of patients (older, sicker, etc.), the results have not been confirmed yet by large prospective, randomized and multicentric investigation. However, if randomized trials on low-risk individuals have already delivered improved results with OPCAB surgery over conventional CABG, one can postulate that the greatest benefit will be in the high-risk group. As the latter group continues to make up a larger percentage of the surgical population in the new millennium, OPCAB surgery may have widespread application. Thus, we, as surgeons, need to make a decision: do we operate on patients with increased Parsonnet risk scores off-pump and low-risk patients on pump, or should we all of them off pump? What is certain is that to perform successful off-pump multivessel coronary surgery in high-risk patients, surgeons must undertake a significant percentage of their total coronary cases without CPB to feel comfortable with high-risk patients. Cardiac surgeons now have access to a new therapeutic tool and the opportunity to get familiar with it. Ultimately they should learn to use it in a scientific manner for the best of their patients.

References

1. Christakis GT, Weisel RD, Fremes SE et al. Coronary artery bypass grafting in patients with poor ventricular function. J Thorac Cardiovasc Surg 1992; 103:1083-92.
2. Elefteriades JA, Tolis G, Levi E et al. Coronary artery bypass grafting in severe left ventricular dysfunction: Excellent survival with improved ejection fraction and functional state. J Am Coll Cardiol 1993; 22:1411-7.
3. Trachiotis GD, Weintraub WS, Johnston TS et al. Coronary artery bypass grafting in patients with advanced left ventricular dysfunction. Ann Thorac Surg 1998; 66:1632-9
4. Angelini GD, Taylor FC, Reeves BC et al. Early and midterm outcome after off-pump and on-pump surgery in Beating Heart Against Cardioplegic Arrest Studies (BHACAS 1 and 2): A pooled analysis of two randomised controlled trials. Lancet 2002; 359:1194-9.
5. Van Dijk D, Nierich AP, Jansen WL et al. Early outcome after off-pump versus on-pump coronary bypass surgery. Results from a randomized study. Circulation 2001; 104:1761-6.
6. Puskas JD, Williams WH, Duke PG et al. Off-pump coronary artery bypass grafting provides complete revascularization with reduced myocardial injury, transfusion requirements, and length of stay: A prospective randomized comparison of two hundred unselected patients undergoing off-pump versus conventional coronary artery bypass grafting. J Thorac Cardiovasc Surg 2003; 125:797-808.
7. Nathoe HM, van Dijk D, Jansen EWL et al. A comparison of on-pump and off-pump coronary bypass surgery in low-risk patients. N Engl J Med 2003; 348:394-402.

8. Arom KV, Flavin TF, Emery RW et al. Is low ejection fraction safe for off pump coronary bypass operation. Ann Thorac Surg 2000; 70:1021-5.

9. Meharwal ZS, Trehan N. Off pump coronary artery bypass grafting in patients with left ventricular dysfunction. Heart Surg Forum 2002; 5(1):41-5.

10. Moshkovitz Y, Sternik L, Paz Y et al. Primary coronary artery bypass grafting without cardiopulmonary bypass in impaired left ventricular function. Ann Thorac Surg 1997; 93:S44-7.

11. Perrault L, Carrier M, Cartier R et al. Morbidity and mortality of reoperation for coronary artery bypass grafting: Significance of atheromatous vein grafts. Can J Cardiol 1991; 7:427-430.

12. Fanning WJ, Kakos GS, Williams TE. Reoperative coronary artery bypass grafting without cardiopulmonary bypass. Ann Thorac Surg 1993; 55:486-9.

13. Trehan N, Mishra YK, Malhotra R et al. Off pump redo coronary artery bypass grafting. Ann Thorac Surg 2000; 70:1026-9.

14. Stamou SC, Pfister AJ, Dangas G et al. Beating heart versus conventional single-vessel reoperative coronary artery bypass. Ann Thorac Surg 2000; 69:1383-1387.

15. Yau TMY, Borger MA, Weisel RD et al. The changing pattern of reoperative coronary surgery trends in 1230 consecutive reoperations. J Thorac Cardiovasc Surg 2000; 120:156-63.

16. Tu JV, Jagal SB, Naylor CD. Multicenter validation of a risk index for mortality, intensive care unit stay, and overall hospital length of stay after cardiac surgery. Steering Committee of the Provincial Adult Cardiac Care Network of Ontario. Circulation 1995; 91:677-84.

17. Ivanov J, Weisel RD, David TE et al. Fifteen-year trends in risk severity and operative mortality in elderly patients undergoing coronary artery bypass graft surgery. Circulation 1998; 97:673-80.

18. Boyd WD, Desai ND, Del Rizzo DF et al. Off-Pump surgery decreases postoperative complications and resource utilization in the elderly. Ann Thorac Surg 1999; 68:1940-3.

19. Stamou SS, Dangas G, Dullum MKC et al. Beating heart surgery in Octogenarians : Perioperative outcome and comparison with younger age groups. Ann Thorac Surg 2000; 69:1140-5.

20. Demers P, Cartier R. Multivessel off pump coronary artery bypass surgery in the elderly. Eur J Cardiothorac Surg 2001; 20(5):908-12.

21. Koutlas TC, Elbeery JR, Williams JM et al. Myocardial revascularization in the elderly using beating heart coronary artery bypass surgery. Ann Thorac Surg 2000; 69:1042-7.

22. Ricci M, Karamanoukian HL, Abraham R et al. Stroke in octogenarians undergoing coronary artery surgery with and without cardiopulmonary bypass. Ann Thorac Surg 2000; 69(5):1471-5.

23. Hoff SJ, Ball SK, Coltharp WH et al. Coronary artery bypass in patients 80 years and over: Is off-pump the operation of choice? Ann Thorac Surg 2002; 74(4):S1340-3.

24. Sabik JF, Gillinov MA, Blackstone EH et al. Does off pump coronary surgery reduce morbidity and mortality? J Thorac Cardiovasc Surg 2002; 124:698-707.

25. Magee MJ, Jablonski KA, Pfister AJ et al. Elimination of cardiopulmonary bypass improves early survival for multivessel coronary artery bypass patients. Ann Thorac Surg 2002; 73(4):1196-202; discussion 1202-3.

26. Cleveland JC, Shroyer ALW, Chen AYC et al. Off-pump coronary artery surgery bypass grafting decreases risk-adjusted mortality and morbidity. Ann Thorac Surg 2001; 72(4):1282-9.

27. Plomondon ME, Cleveland Jr JC, Ludwig ST et al. Off-pump coronary artery bypass is associated with improved risk-adjusted outcomes. Ann Thorac Surg 2001; 72(1):114-9.

28. Al-Ruzzeh S, Ambler G, Asimakopoulos G et al. Off pump coronary artery bypass surgery reduces morbidity and risk stratified mortality. Local and national comparisons: Initial UK multi-centre analysis of early clinical outcome. Circulation 2003; 108(Suppl 1):II1-8.

29. Bitner HB, Savitt MA. Off pump coronary artery bypass grafting decreases morbidity and mortality in a selected group of high risk patients. Ann Thorac Surg 2002; 74 :115-8.

30. Carrier M, Perrault LP, Page P et al. Randomized trial comparing off-pump to on-pump cabg in high risk patients. Heart Surg Forum 2003; 6(Supp 1):S14-20.

31. Golding LR, Loop FD, Phillips D et al. Use of intra-aortic balloon pump in cardiac surgical patients; the Cleveland Clinic experience, 1975-1976. Cleve Clin Q 1976; 43(3):117-20.

32. Juhlin-Dannfelt A, Nordlander R, Nyquist O. Peripheral hemodynamics in assisted circulation with intra-aortic balloon pumping in patients with cardiogenic shock. Acta Med Scand 1979; 205(6):505-8.

33. Kim KB, Lim C, Ahn H et al. Intraaortic balloon pump therapy facilitates posterior vessel off pump coronary artery bypass grafting in high risk patients. Ann Thorac Surg 2001; 71:1964-8.
34. Craver JM Murrah PC. Elective intraaortic balloon pump counterpulsation for high risk off pump coronary artery bypass operations. Ann Thorac Surg 2001; 71:1220-3.
35. Babatasi G, Massetti M, Bruno P et al. Preoperative balloon counterpulsation and off-pump coronary surgery for high-risk patients. Cardiovasc Surg 2003; 11(2):145-8.
36. Kolesov VI. Mammary artery-coronary artery anastomosis as method of treatment for angina pectoris. J Thorac Cardiovasc Surg 1967; 54(4):535-44.
37. Benetti FJ, Mariani MA, Ballester C. Direct coronary surgery without cardiopulmonary bypass in acute myocardial infarction. J Cardiovasc Surg (Torino) 1996; 37:391-5.
38. Wos S, Jasinski MJ, Bachowski R et al. Is primary off-pump CABG an option in acute myocardial infarction. Heart Surg Forum 2003; 6(Suppl):S42.
39. Ochi M, Hatori N, Saji Y et al. Application of off-pump coronary artery bypass grafting for patients with acute coronary syndrome requiring emergency surgery. Ann Thorac Cardiovasc Surg 2003; 9:29-35.

CHAPTER 12

Pitfalls in Off-Pump Coronary Artery Bypass Surgery

Nicolas Dürrleman and Raymond Cartier

Introduction

Through the last 5 years off-pump coronary artery bypass (OPCAB) surgery has constantly gained popularity in the cardiovascular community. Now accounting for 20% of the coronary artery bypass grafting (CABG) practice in the U.S.A. and probably up to 15% in Canada, its use is expected to grow.[1] As any other novel technique, OPCAB surgery has its drawbacks and limitations. Not manifest at the beginning, these drawbacks progressively catch-up with the operator. Recognizing the obstacles should not mean defeat, but rather wisdom and understanding. Then, sharing them with others becomes a matter of honesty and responsibility. Many authors have reported their experience and learning curves in off-pump surgery.[2-6] In this chapter, we join them in discussing our work in the area and underscore the major pitfalls and drawbacks that we have encountered over our 8 year experience with OPCAB surgery at the Montreal Heart Institute.

The Enlarged Aorta (Fig. 1)

With constant aging of the current surgical population, degenerative atherosclerotic changes of the ascending aorta are common findings. In these circumstances, side-clamping of the physiological, pressurized aorta is troublesome and can predispose to aortic dissection.[7] Early in our institutional experience, we were confronted with patients who sustained ascending aortic dissection (AAD) due to side-bite clamping, which raised concerns about the incidence of this complication. With time and experience however, its incidence has gone down. Currently, in our practice, over 900 systematic OPCABs, acute AADs in the first week of the procedure have occurred on 3 occasions (0.3%). In the first case, the dissection was recognized at surgery, and the patient underwent successful ascending aortic replacement. In the two other patients, the dissection occurred 1 week after the initial procedure. Both patients died on the table during a surgical attempt to repair the ascending aorta.

Early in our experience, we were using the "Beck clamp", which has sharp edges and pointy teeth that grab the aortic wall quite strongly and occasionally damage it. Since these events, we have changed to a soft clamp with padded arms (Fig. 2) that are less damaging to the aorta. We religiously lower systolic blood pressure before and during the aortic side-clamping to 90-100 mmHg. Interestingly, all patients who presented aortic dissection had clinical and histological evidence of medial disease. We currently avoid side-clamping moderately-enlarged aortas ($\cong$ 4 cm diameter) and favour, in these circumstances, arterial revascularization with side-branch grafts to avoid aortic manipulation. To alleviate the need for partial clamping of

Off Pump Coronary Artery Bypass Surgery, edited by Raymond Cartier. ©2005 Eurekah.com.

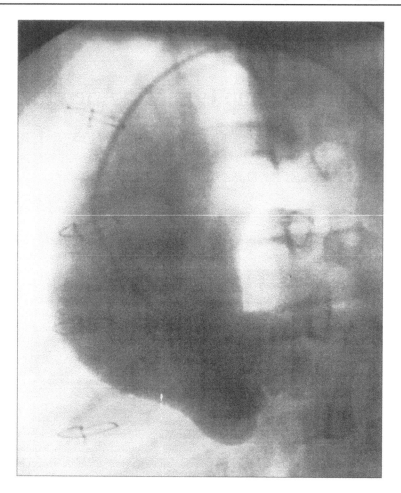

Figure 1. Enlarged aorta in an elderly patient. See text for explanation.

the ascending aorta other options are available. The new device named "Enclose device" (Novare, Cupertino, CA, USA) was found particularly useful to create an enclosed bloodless field approximately one centimeter square in the fully pressurized aorta eliminating the need for side clamping (Fig. 2B). The surgeon can construct a proximal end-to-side anastomosis using standard suturing technique. Multiple anastomoses can be completed with the same device. We have used this device in many cases where the fragility of the ascending aorta precluded conventional manipulation of the aorta with success.

Aortic dissection is not limited to OPCAB surgery, and was already reported during the early days of "on-pump" cardiac surgery.[8] Ruchat and colleagues, in a review of 8,624 patients who underwent cardiac surgery with cardiopulmonary bypass (CPB) and cardioplegic arrest, identified 10 cases (0.12%) with AAD.[9] Among them, 7 dissected intra-operatively and the remaining 3, 7-38 days after surgery. This prevalence is still lower than what we have reported (0.3%) but could have been higher if only CABG surgery had been considered with side-clamping of the aorta. Nevertheless, AAD is a complication that any "OPCABers" should keep in mind during off-pump surgery. Surgeons should not hesitate to modify their strategy when aortic friability is suspected.[10]

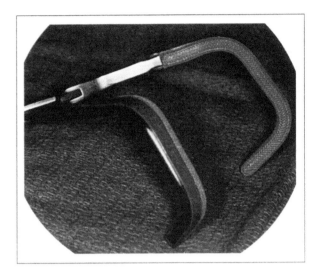

Figure 2A. Soft-padded aortic side-clamp.

The Occluded Vessel

The occluded target vessel could be a source of problems (Fig. 3). Classically, in our experience, these vessels have thick walls and small residual lumen. Proceeding to arteriotomy without circulating blood inflow increases the risk of injury and occasionally leads to wall dissection. Frequently an adjunctive coronary endarterectomy is necessary, which can further augment morbidity from the procedures.[11,12] Most likely, the best strategy to prevent this problem is to avoid any blood flow interruption during the opening. Pressurized blood flowing through the initial arteriotomy allows quicker identification of the endovascular lumen and further enlargement without jeopardizing its integrity.

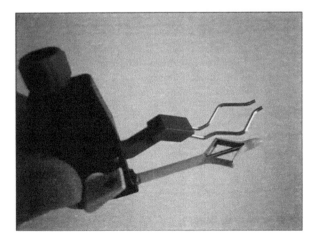

Figure 2B. The "Enclose device" from Novare.

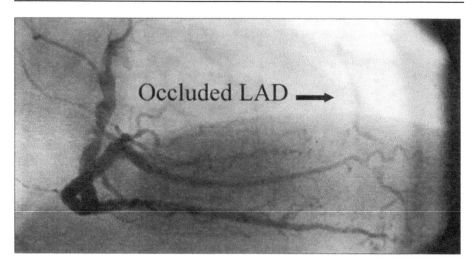

Figure 3. Angiogram showing an occluded LAD.

The Dominant Right Coronary Artery (RCA)

Classically, the RCA network has a dominant feature in 85-90% of the normal anatomy.[13] Typically, it is the "neglected vessel" in conventional cardiac surgery, as it is generally grafted at the beginning of the procedure and then forgotten. Classically, the RCA crosses the crux and takes the shape of a deep U turn. The atrioventricular node artery arises at the apex of the "U".[14] The atrioventricular artery is usually supplied by the dominant RCA. In cases of non-critical RCA stenosis, acute occlusion could cause a temporary ischemic atrioventricular block. This can be problematic if it occurs during grafting of the RCA. Few options are available to avoid severe bradycardia or atrioventricular bloc. The surgeon can directly graft the posterior descending artery or use a shunt during the anastomosis (at least 2 mm diameter). Another option is to deploy temporary epicardial or endovascular pacing at the time of the anastomosis.

The Dominant Left Main (Fig. 4)

Critical disease of a dominant left main could be quite challenging if the RCA network is nondominant or nonreconstructable. Blood flow interruption in the left anterior descending artery (LAD) territory is generally poorly tolerated in this set-up because neither the circumflex nor the posterior descending artery has enough circulating blood flow to compensate. Shunting of the LAD becomes vital to minimize life-threatening ischemia, and should be inculcated without hesitation.

Myocardial ischemia is predictable in some situations. A noncritical RCA stenosis is one of them as discussed previously. Other circumstances are the noncritical stenosis of the LAD or a generous diagonal artery. A 2 minute preischemic test generally rules out or identifies potentially life-threatening ischemia.

Ischemic Mitral Insufficiency (IMI)

The management of concomitant IMI at the time of CABG has been and remains a controversial topic.[15,16] Persistence of residual mitral insufficiency is associated with decreased long-term survival. The current trend favours the combined repair of any significant IMI (>2+)

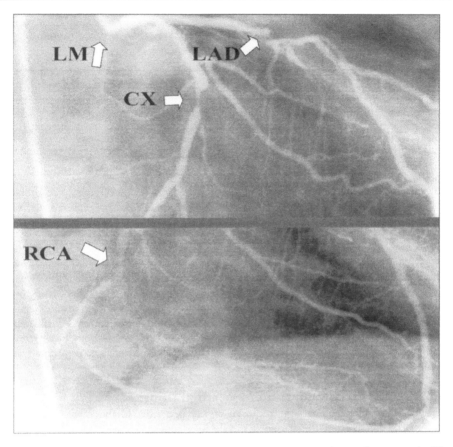

Figure 4. Angiogram of a patient with a left dominant coronary artery. Note the significant stenosis of the left main artery, and the significant disease involving the anterior, inferior, and posterior territories. See text for explanation. LM= left main artery; LAD= left anterior descending artery; CX= circumflex artery, RCA= right coronary artery.

at the time of CABG in a single-stage procedure; but even with excellent immediate results, an early recurrent risk of up to 30% still prevails. According to Tahta[17] and colleagues, a decline of left ventricular function and a medical history of myocardial infarction prior to surgery are predisposing factors for early recurrence. Nonrecognition of IMI is common in conventional surgery.[18] Under general anesthesia, severe (grade IV) IMI can be significantly downgraded to nonsignificant grade 0-II in a majority of patients by intraoperative transesophageal echocardiographic evaluation, leaving the surgeon in a puzzling situation.[18] On the other hand, it can rarely be neglected during off-pump surgery. In general, patients with significant preoperative IMI poorly tolerate the surgical manipulation and mobilization that could be further enhanced by temporary ischemia. The first question that should be asked in front of a hemodynamically unstable patient during OPCAB surgery is the functional state of the mitral valve. Transesophageal echocardiography is a primordial tool to rule out this diagnosis and should be routinely used in patients with decreased left ventricular function or a history of recent cardiac failure. Any significant IMI should preclude the OPCAB option, but ultimately, the decision to repair the valve should be finalized preoperatively to avoid any unnecessary attempt at off-pump surgery.

Table 1. Angiographic patency and OPCAB surgery according to the Fitzgibbon classification

Authors (n)	Patency	"A"	"B"	"0"
Nathoe (148)*	91% (93%)			
Puskas (30)*				
Puskas (167)	98.8%	93.3%	5.5%	1.2%
Nakajima (218)	98.5%			1.5%
Kim (223)	99.4%			
Corbineau (42)	98.1%	90.5%	7.6%	1.9%
Nakamura (61)	97.0%	87.4%	9.6%	3.0%
Kobayashi (238)	97.8%	88.6%	9.2%	2.2%

*Prospective, randomized studies

Occasionally, de novo mitral insufficiency occurs during OPCAB surgical manipulations. This is generally associated with elevated pulmonary artery pressure (PAP) and hemodynamic perturbations. As described in Chapter 6, these can be managed by temporary snaring of the inferior vena cava. The manoeuvre instantly decreases cardiac preload and restores left ventricular dynamics. It can be adjusted (with the snare) to slowly normalize the PAP and kept in place for a few minutes. As PAP resumes physiological values the snare is progressively released. This gives the surgeon time to safely finish the grafting without having to perform an emergency conversion.

Mobilization Techniques and Myocardial Injuries

As discussed in Chapter 6, the anchoring of deep pericardial stitches can cause problems. Injuries to the descending thoracic aorta and the lung parenchyma with severe bleeding have been reported but are isolated cases.[19-21] Pericardial stitches have to be inserted superficially and under direct vision. Apical suction reduces risk of posterior bleeding but can cause abrasions on the myocardium and occasionally, regional ischemia by interfering with local arterial perfusion.[22] The key to avoiding these complications is to be aware of the technical pitfalls, and keep them in mind when performing such techniques.

Graft Patency

Graft patency after OPCAB surgery remains a controversial and much-debated subject. To date, a number of randomized and nonrandomized studies have thoughtfully examined the question. Table 1 summarizes the most recent publications on the topic. Two recent prospective and randomized studies have reported similar early graft patency with OPCAB and on-pump CABG.[23,24] Nathoe et al[23] in a randomized, prospective investigation, recorded patencies of 91% and 93% for off- and on-pump techniques. Puskas et al[24-25] also found similar graft patency in a small subset of diabetic patients randomized between on- and off-pump CABG. They and others, have presented excellent results in nonrandomized studies.[25-30] Most of them have claimed global patency rates exceeding 98% with at least 90% good results (grade A Fitzgibbon).[31] Nakamura et al[32] obtained a 10-15% rate of nonfunctional grafts ("string sign" or blood flow "steel") with composite conduits made of internal thoracic artery (ITA) and radial arteries, but their global patency rate with complete arterial revascularization was high. Stenosis below 75% in the native artery and the use of more than 4 sequential grafts were evocated as risk factors for competitive flow.[29,32] For comparative purposes, the patency rates at

33 months and 7 years obtained by Bakay et al[33] and Dion et al[34] with conventional coronary surgery were 91% and 96% respectively with ITA sequential bypasses performed under CPB. This is quite comparable to patency rate reported with OPCAB surgery.[33,34]

Hemodynamic Instability and Acute Conversion

Acute conversion to CPB is the most feared complication in OPCAB surgery. In the early days, conversion rates were as high as 25%.[35] However, in more recent literature, the rate varies between 0 to 9.2%.[36-40] Mathur, reporting on 124 cases in a community hospital, considered the complexity level of the revascularization procedure as being the principal risk-factor for conversion.[36] The 5.6% incidence of acute conversion occurred in "complex multi-vessel OPCAB" consisting mainly of patients presenting triple vessel disease necessitating 3 or more bypass grafts. The post-operative evolution of these "converted cases" was not as good as the others mainly because of increased morbidity and decreased graft permeability. He concluded that in his practice the use of off-pump surgery was now limited to single or double vessel disease.

Mujanovic and colleagues reported a conversion rate of 9.2%.[37] Among the 36 patients they have to convert (out of 357 OPCAB procedures), 21 were for hemodynamic instability, 8 for graft revision, 5 for ventricular fibrillation, and 2 for poor native runoff. The outcome was poor among 9 patients who had been seriously unstable or had ventricular fibrillation. In these cases, the mortality rate reached 66% (6/9). They concluded that myocardial ischemia had to be carefully monitored and treated aggressively during surgery, and the timing of conversion was crucial to avoid serious outcomes. Other authors have been less pessimistic and have noted a lower incidence of conversion with favourable outcome.[38-40] Vassiliades and colleagues have reported 23 cases of acute conversion (1.6%) in a series of 1,420 OPCAB surgeries.[39] The majority of them (20/23) had an uneventful post-op evolution. According to these authors, conversions were due to ischemic or mechanical causes, and, most of the time, to judgemental errors or intra-operative technical mistakes, suggesting that with experience these conversions could be avoided.

Compression of the right ventricle is also an important cause of instability and potential conversion as reported by Hart.[40] Myocardial mobilization always evokes to some extent, restrictive diastolic disease of the right ventricle. This could be attenuated by herniation of the right ventricle into the right pleural space, as discussed in Chapter 6. Although the experimental use of apical suction was demonstrated to improve hemodynamic stability during verticalization in the porcine model,[41] a recent randomized study in humans did not reveal any specific advantage over classical deep pericardial stitches.[42]

It appears that with experience, patience and vigilance, the rate of conversion can be reduced and kept quite low. In our experience with 940 consecutive and systematic cases, we experimented on 4 conversions, 1 for ventricular fibrillation, 1 for unstable hemodynamics, 1 for a deep intramyocardial LAD, and 1 for acute AAD.[43] Their evolution was uneventful. We still keep an unprimed extracorporeal circuit "ready for use" in the operating theatre, with a perfusionist on standby.

Post-Operative Thromboembolic Complications

Many authors have expressed concern about a potential state of hypercoagulability following OPCAB surgery.[44-47] It has even been suggested that the patency of saphenous vein grafts (but not arterial conduits) is reduced in OPCAB patients.[48] Full heparinization is not a routine in OPCAB surgery, and the usual post-operative "coagulopathy" observed in conventional surgery is theoretically avoided. In a small cohort of minimally invasive direct coronary artery bypass and OPCAB patients, Mariani and colleagues noted that pro-coagulant activity was expressed by an elevated level of prothrombin factors 1 and 2 along with increased fibrinolysis and von Willebrand factor.[44] This "pro-coagulable state" was temporary, lasting only the first post-operative day, and not influencing the post-operative course. In a small nonrandomized series, Quigley and colleagues[46] found that an heightened coagulation index (evaluated by thromboelastography) during the first 3 days after OPCAB surgery was not present in conventional CABG patients. Others have discovered increased platelet activation in the first 24 hours after OPCAB, but not in conventional CABG patients.[47] All these observations have led some authors to reconsider the optimal anti-platelet regimen after OPCAB.[49] Kurlanski[49] in a recent editorial, has suggested that perhaps Clopidogrel should be administered instead of aspirin as the post-operative antiplatelet regimen. These studies could not demonstrate any specific clinical effect of this "post-operative hypercoagulable state" on graft patency and did not give any consideration to possible venous thromboembolic complication.

In a previous publication, we have reported an increased incidence of deep venous thrombosis and pulmonary embolism after OPCAB surgery.[50] Compared to conventional surgery in our institution, the incidence was not statistically significant (1% vs 0.5%). However, most of these complications occurred in "healthy" OPCAB patients whereas patients operated on with CPB had more comorbid factors for thromboembolic venous disease. Since then, we have modified our anti-platelet regimen so that all patient are maintained on aspirin (325 mg) started on the day of surgery, Clopidogrel (75 mg) started on day 1 for 30 days, and heparin (5000 units TID) s-c during the entire hospitalization period. We have noticed a reduced incidence of venous thromboembolic complications among our patients.

Training

The ability to train residents to perform OPCAB surgery has been questioned in diverse institutions devoted to teaching. Essentially, can we or can we not educate young surgeons without specific skills in coronary artery surgery to perform OPCAB surgery safely? This question has been examined by a few institutions and deserves consideration.[50-53] Karamanoukian and his group[51] from the State University of New York at Buffalo recommended that every trainee perform at least 50 cases under the supervision of a trained staff surgeon to become familiar with the procedure. This was based on the successful experience of a single resident. Caputo et al[52] from the Bristol group have twice reported their experience with training in off-pump surgery.[52,53] In 2001, already 40% of their CABG case load was carried out by trainees, 44% of the surgeries being off-pump. Early and mid-term outcomes were comparable for off- and on-pump patients who were operated on by either consultants or trainees. They did not mention the number of trainees who were involved in this review. Jenkins et al[54] from Middlesex, U.K., reported the experience of 4 trainees with OPCAB surgery in 64 patients. Circumflex territory was grafted in 75% of the cases. Coronary angiographies performed on 5 patients revealed 100% graft patency. They concluded that with appropriate training, trainees could learn to perform successful multi-vessel revascularization in relatively high-risk patients. Clearly teaching OPCAB surgery is a demanding task for senior consultants. However, it could be taught as any other technique as long as the teacher has solid experience, feels confident with it, and has the desire to share it with trainees.

Comments

During the last decade, off pump coronary surgery has come a long way. From a single bypass to the anterior wall of the heart without stabilization to complex multiple bypass grafting aided by stabilizers, surgeons have shown how to use the versatility of their imagination. As they brought OPCAB surgery forward, they learned to deal with its various limits and drawbacks. In a certain way, acknowledging the limitations of the procedure is learning how to use it in best efficient manner. Clinically, this leads to maximizing benefits and minimizing side effects on the patients.

One can always say that all these difficulties can be easily overcome with conventional on-pump surgery. Over the years though, new technologic advancement in interventional cardiology has set back the coronary surgery to the second line. Surgeons are now confronted with more complex surgery affecting older and sicker patients. Adding to this new reality, the economic pressure of the health care system where accessibility and efficiency at best possible cost is sought for, the incentive to move on toward less invasive "as efficient" procedure is strong.

To perform new procedures is not always an easy task. A frequently asked question is how many procedures a surgeon should perform before being comfortable with it. In the particular setup of OPCAB surgery, the patients that will profit the most of a less invasive procedure are not always the easiest ones to deal technically. Co-morbid factors such as ventricular dysfunction, renal failure, calcified aorta, small or calcified vessels, etc… are frequently seen in these patients. To proceed off-pump in those patients requires a great deal of experience. It is then imperative that surgeons get their hands on easy cases to learn how to master the procedure.

Any time a new technology is introduced in clinical practice the issue of quality control has to be addressed. Monitoring of complications and early angiographic controls on the first patients are primordial for objective evaluation of the surgical performance. The surgeon can readjust his technique if problems occur and reorient the indications of the procedure. Progressively, with experience, expertise and confidence increase. In our current era of rapid technical evolution, surgeons are and will be constantly challenged with new procedures. They have to be prepared to meet these new challenges and live up to their expectations.

References

1. Mack MJ. Coronary surgery: Off-pump and port access. Surg Clin North Am 2000; 80(5):1575-91.
2. Hart JC. Hemodynamic stability and lessons learned: A four-year OPCAB experience. Heart Surg Forum 2001; 4(4):335-8.
3. Zehr KJ, Handa N, Bonilla LF et al. Pitfalls and results of immediate angiography after off-pump coronary artery bypass grafting. Heart Surg Forum 2000; 3(4):293-9.
4. Bergsland J, D'Ancona G, Karamanoukian H et al. Technical tips and pitfalls in OPCAB surgery: The Buffalo Experience. Heart Surg Forum 2000; 3(3):189-93
5. Vassiliades Jr TA, Nielsen JL, Lonquist JL. Hemodynamic collapse during off-pump coronary artery bypass grafting. Ann Thorac Surg Jun 2002; 73(6):1874-9; discussion 1879.
6. Hart JC. Maintaining hemodynamic stability and myocardial performance during off-pump coronary bypass surgery. Ann Thorac Surg Feb 2003; 75(2):S740-4.
7. Chavanon O, Carrier M, Cartier R et al. Increased incidence of acute ascending aortic dissection with off-pump aortocoronary bypass surgery? Ann Thorac Surg Jan 2001; 71(1):117-21.
8. Boruchow IB, Iyengar R, Jude JR. Injury to ascending aorta by partial-occlusion clamp during aorta-coronary bypass. J Thorac Cardiovasc Surg 1977; 73(2):303-5.
9. Ruchat P, Hurni M, Stumpe F et al. Acute ascending aortic dissection complicating open heart surgery: Cerebral perfusion defines the outcome. Eur J Cardiothorac Surg 1998; 14(5):449-52.
10. Hagl C, Ergin MA, Galla JD et al. Delayed chronic type A dissection following CABG: Implications for evolving techniques of revascularization. J Card Surg 2000; 15(5):362-7.
11. Djalilian AR, Shumway SJ. Adjunctive coronary endarterectomy: Improved safety in modern cardiac surgery. Ann Thorac Surg 1995; 60(6):1749-54.
12. Qureshi SA, Halim MA, Pillai R et al. Endarterectomy of the left coronary system. Analysis of a 10 year experience. J Thorac Cardiovasc Surg 1985; 89(6):852-9.
13. James TN. Anatomy of the Coronary Arteries. Hoeber, New York: 1961.
14. Soto B, Russell RO, Moraski RE. Radiographic Anatomy of the Coronary Arteries: An Atlas. Futura, Mount Kisco: 1976.
15. Adams DH, Filsouf F, Aklog L. Surgical treatment of the ischemic mitral valve. J Heart Valve Dis 2002; 11(Suppl 1):S21-5.
16. Tolis Jr GA, Korkolis DP, Kopf GS et al. Revascularization alone (without mitral valve repair) suffices in patients with advanced ischemic cardiomyopathy and mild-to-moderate mitral regurgitation. Ann Thorac Surg 2002; 74(5):1476-80.

17. Tahta SA, Oury JH, Maxwell JM et al. Outcome after mitral valve repair for functional ischemic regurgitation. J Heart Valve Dis 2002; 11(1):11-8.

18. Aklog L, Fisoufi F, Chen RH et al. Does coronary artery bypass grafting alone correct moderate ischemic mitral regurgitation? Circulation 2001; 104(12Suppl1):168-75.

19. Salerno TA. Letter to the Editor: A word of caution on deep pericardial sutures for off-pump coronary surgery bypass procedures. Ann Thorac Surg 2003; 76:339.

20. Fukui T, Suehiro S, Shibata T et al. Retropericardial hematoma complicating off-pump coronary artery bypass surgery. Ann Thorac Surg 2002; 73(5):1629-31.

21. Zamvar V, Deglurkar I, Abdullah F et al. Bleeding from the lung surface: A unique complication of off-pump CABG operation. Heart Surg Forum 2001; 4(2):172-3.

22. D'Ancona G, Karamanoukian H, Kawaguchi A et al. Coronary artery exposure in off-pump CABG: A word of caution. Heart Surg Forum 2001; 4(3):243-5; discussion 245-6.

23. Nathoe HM, van Dijk D, Jansen EW et al. A comparison of on-pump and off-pump coronary bypass surgery in low-risk patients. N Engl J Med 2003; 348(5):394-402.

24. Puskas JD, Sharoni E, Petersen R et al. Angiographic graft patency and clinical outcomes among diabetic patients after off-pump versus conventional coronary artery bypass grafting: Results of a prospective randomized trial. Heart Surg Forum 2003; 6(Suppl 1):S27.

25. Puskas JD, Thourani VH, Marshall JJ et al. Clinical outcomes, angiographic patency, and resource utilization in 200 consecutive off-pump coronary bypass patients. Ann Thorac Surg 2001; 71(5):1477-83; discussion 1483-4.

26. Nakajima H, Kobayashi J, Tagusari O et al. Competitive flow in arterial composite grafts and effect of graft arrangement in off-pump coronary revascularization. Heart Surg Forum 2003; 6 (Suppl 1):S14.

27. Kim KB, Cho KR, Chang WI et al. Bilateral skeletonized internal thoracic artery graftings in off-pump coronary artery bypass: Early result of Y versus in situ grafts. Ann Thorac Surg 2002; 74(4):S1371-6.

28. Corbineau H, Verhoye JP, Langanay T et al. Feasibility of the utilisation of the right internal thoracic artery in the transverse sinus in off pump coronary revascularisation: Early angiographic results. Eur J Cardiothorac Surg 2001; 20:918-22.

29. Nakamura Y, Kobayashi J, Tagusari O et al. Early results of complete off-pump coronary revascularization using left internal thoracic artery with composite radial artery. Jpn J Thorac Cardiovasc Surg 2003; 51(1):10-5.

30. Kobayashi J, Tagusari O, Bando K et al. Total arterial off-pump coronary revascularization with only internal thoracic artery and composite radial artery grafts. Heart Surg Forum 2002; 6(1):30-7.

31. Fitzgibbon GM, Kafka HP, Leach AJ et al. Coronary bypass graft fate and patient outcome: Angiographic follow-up of 5,065 grafts related to survival and reoperation in 1,388 patients during 25 years. J Am Coll Cardiol 1996; 28:616-26.

32. Nakajima H, Kobayashi J, Tagusari O et al. Competitive flow in arterial composite grafts and effect of graft arrangement in off-pump coronary revascularization. Heart Surg Forum 2003; 6(Suppl 1):S14.

33. Bakay C, Erek E, Salihoglu E et al. Sequential use of internal thoracic artery in myocardial revascularization: Mid- and long-term results of 430 patients. Cardiovasc Surg 2002; 10(5):481-8.

34. Dion R, Glineur D, Derouck D et al. Long-term clinical and angiographic follow-up of sequential internal thoracic artery grafting. Eur J Cardiothorac Surg 2000; 17(4):407-14.

35. Soltoski P, Salerno T, Levinsky L et al. Conversion to cardiopulmonary bypass in off-pump coronary artery bypass grafting: Its effect on outcome. J Card Surg 1998; 13(5):328-34.

36. Mathur AN. Hemodynamic collapse during off-pump coronary artery bypass and a review of some controversial adverse outcomes. Heart Surg Forum 2003; 6(Suppl 1):S52.

37. Mujanovic E, Kabil E, Hadziselimovic M et al. Conversions in off-pump coronary surgery. Heart Surg Forum 2003; 6(3):135-7.

38. Anyanwu AC, Al-Ruzzeh S, George SJ et al. Conversion to off-pump coronary bypass without increased morbidity or change in practice. Ann Thorac Surg 2002; 73(3):798-802.

39. Vassiliades Jr TA, Nielsen JL, Lonquist JL. Hemodynamic collapse during off-pump coronary artery bypass grafting. Ann Thorac Surg 2002; 73(6):1874-9; discussion 1879.

40. Hart JC. Maintaining hemodynamic stability and myocardial performance during off-pump coronary bypass surgery. Ann Thorac Surg 2003; 75(2):S740-4.
41. Sepic J, Wee JO, Soltesz EG et al. Cardiac positioning using an apical suction device maintains beating heart hemodynamics. Heart Surg Forum 2002; 5(3):279-84.
42. Gummert JF, Raumanns J, Bossert T et al. Suction device versus pericardial retraction sutures. Comparison of hemodynamics using different exposure systems in beating heart coronary surgery. Heart Surg Forum 2003; 6(Suppl 1):S32.
43. Cartier R. Current trends and technique in OPCAB surgery. J Card Surg 2003; 18(1):32-46.
44. Mariani MA, Gu YJ, Boonstra PW et al. Procoagulant activity after off-pump coronary operation: Is the current anticoagulation adequate? Ann Thorac Surg 1999; 67(5):1370-5.
45. Casati V, Gerli C, Franco A et al. Activation of coagulation and fibrinolysis during coronary surgery: On-pump versus off-pump techniques. Anesthesiology 2001; 95(5):1103-9.
46. Quigley RL, Fried DW, Pym J et al. Off-pump coronary artery bypass surgery may produce a hypercoagulable patient. Heart Surg Forum 2003; 6(2):94-8.
47. Moller CH, Steinbruchel DA. Platelet function after coronary artery bypass grafting: Is there a procoagulant activity after off-pump compared with on-pump surgery? Scand Cardiovasc J. Jul 2003; 37(3):149-53.
48. Kim KB, Lim C, Lee C et al. Off-pump coronary artery bypass may decrease the patency of saphenous vein grafts. Ann Thorac Surg 2001; 72(3):S1033-7.
49. Kurlansky PA. Is there a hypercoagulable state after off-pump coronary artery bypass surgery? What do we know and what can we do? J Thorac Cardiovasc Surg 2003; 126(1):7-10.
50. Cartier R, Robitaille D. Thrombotic complications in beating heart surgery. J Thorac Cardiovasc Surg 2002; 121:920-2.
51. Karamanoukian HL, Panos AL, Bergsland J et al. Perspectives of a cardiac surgery resident in-training on off-pump coronary bypass operation. Ann Thorac Surg 2000; 69(1):42-5; discussion 45-6.
52. Caputo M, Chamberlain MH, Ozalp F et al. Off-pump coronary operations can be safely taught to cardiothoracic trainees. Ann Thorac Surg 2001; 71(4):1215-9.
53. Caputo M, Bryan AJ, Capoun R et al. The evolution of training in off-pump coronary surgery in a single institution. Ann Thorac Surg 2002; 74(4):S1403-7.
54. Jenkins D, Al-Ruzzeh S, Khan S et al. Multi-vessel off-pump coronary artery bypass grafting can be taught to trainee surgeons. Heart Surg Forum 2002; 5 (Suppl 4):S342-54.

CHAPTER 13

Sutureless Coronary Artery Bypass Grafting:
Experimental and Clinical Progress

Kenton J. Zehr

Historical Perspective and Impetus for the Development of Mechanical Anastomoses

The vision of vascular surgery became clearer in 1902, when Alexis Carrel described his triangulation technique for suturing blood vessels together.[1] He experimented with vessels of many sizes and thickness. Suturing was able to accommodate these discrepancies, and the technique allowed for patent, reproducible anastomoses. Originally, sutures were thick and braided. Relatively inert, monofilament, fine suture was developed. This effectively eliminated intravascular thrombus formation. Anastomoses of very small vessels with high patency rates became possible. Further advancements have been made with the development of surgical loupes and operative microscopes for magnification, topical hemostatic agents, including glues and fibrin products, artificial conduits and repair patches, endovascular stents for intra-luminal approaches to revascularization, and the ability to provide surgical control of the results, using doppler ultrasound, thermal and contrast angiography. Surgeons now perform a wide variety of advanced surgical procedures, such as experimental and clinical transplantation of organs, digit and extremity re-attachments, free flap composite tissue reconstruction, and direct coronary artery revascularization. In the last one hundred years, surgical subspecialties have evolved different techniques for creating vascular anastomoses. However, the common thread to all of these procedures remains a "thread".

As more sub-specialties working across the globe advanced their specific techniques for vascular anastomoses, work directed toward facilitated anastomoses in one sub-specialty or country may have gone unnoticed elsewhere. For example, Androsov,[2] working in Moscow in the 1950s, made a device that facilitated vascular connections by creating circular U-clip anastomoses, a type of semi-automated stapling. He performed 354 procedures in dogs with minimal thrombosis or pseudointimal hyperplasia problems. Subsequently, the technique was adapted for limb re-attachment and excision of aneurysms with primary re-anastomoses in a large number of patients with good results. Despite success and ultimate publication in the English literature, most surgeons were unaware of these advancements. The device became obsolete largely due to its relative unavailability. Its use was limited by the requirement of soft distensible vessels for the necessary 180° eversion. Little research and development were continued in the area. As in much of life and surgical technique, Androsov's device was reinvented by several groups in the following decades,[3-6] most without corroboration from each other.

Interest in connecting vessels with coupling rings coincided with the development of suture techniques. Payr[7] published his results in 1904 with an absorbable magnesium ring for end-to-end anastomoses. Most coupling devices provoked necrosis of the vessel due to ischemia related to the compression they caused. A flurry of experiments ensued with variants of this technique involving perforated rings and pins in the 1950s and 1960s.[8-12] These devices were deployed in early experimental and clinical direct coronary artery revascularization procedures. Despite their successes, individual gains went largely unnoticed, and further advancements with specific devices did not occur.

Most work on the development of mechanical devices occurred alongside the clinical advances in general surgery in the last 50 years. Stapling devices were adapted for bowel anastomoses. The results were rapid, uniform, and reproducible. They became standards of care, in contrast to hand-sewing methods. Attempts were made to adapt the devices to vascular anastomoses. However, they proved cumbersome, did not handle size variations well, were unable to deal with atherosclerotic vessels, or would not work with mm-sized or smaller vessels. In most vascular cases, the required number of anastomoses was usually 1 or 2, and variations in size and vessel quality were extreme, i.e., from aorta to dorsalis pedis. The procedure was not rushed. A good surgeon with sutures was able to adjust for these discrepancies and take the time to perform a perfect anastomosis in a bloodless, still field. This ability became the art of vascular surgery.

Direct coronary artery bypass grafting (CABG) developed in the experimental laboratories of Demikhov,[13] and Murray[14] during the 1950s prior to the introduction of cardiopulmonary bypass (CPB). Because of the inability to obtain a still, bloodless field due to the inability to stop the heart for the time it would take to suture an anastomosis, the experimenters employed coupling devices to perform these anastomosis rapidly. Buoyed by this experience, Goetz accomplished a right internal thoracic artery (ITA) to right CABG in 17 seconds on May 2, 1960, using a perforated tantelum ring in a modification of Payr's technique.[15] This was the first direct coronary artery bypass performed clinically. Kolesov followed suit, with a suture technique in 1964, and then with a stapling method in 1967 to anastomose the left ITA to the left anterior coronary artery in a small series of patients.[16] Although the initial experimental and clinical coronary revascularization was performed mechanically, suturing the anastomosis became the preferred approach. The reasons were similar to previous experience with peripheral vascular surgery. An adequate device to perform an end-to-side anastomosis into a diseased vessel did not exist. There was no longer an impetus to continue developing such a device with the introduction of CPB. The ability to do surgery in a bloodless, still field brought the procedure back into the comfort zone of surgeons. Rapid progress was made in surgical technique to allow from 5-10 anastomoses within the safe time for CPB. A legion of cardiac surgeons was trained with this system. The introduction of CPB and cardioplegia resulted in a simplification of direct CABG, making it technically easier.

Conventional CABG is an intensive effort. Advanced technology is required for perfusion and oxygenation. A substantial economic investment is necessary to start and maintain the system and the cases are expensive. Coronary revascularization is an epicardial procedure. Theoretically, it can be performed without stopping the heart. While advancements in the conventional approach were the introduction of ITA grafts and other arterial conduits, several centres continued to develop the technique of performing direct coronary artery revascularization on the beating heart.[17-19] The primary reason was to avoid the expense of the CPB system. The most significant advance was the discovery of manipulations to position the heart and approach all its areas. Additional ideas of shunting blood within the coronary territory while performing the anastomosis and the development of stabilization devices to decrease motion in the operative area resulted in a relatively safe procedure.

There has been global interest in off-pump revascularization in the last 5 years. It is estimated that 15-20% of CABG is done without the use of CPB. In some situations, the driving force is economic benefit. In others, interest seems to be driven by the fact that it can be done.

The prospect of avoiding the small but present risk of complications from CPB is the most attractive benefit of off-pump coronary surgery. Clearly, it has been shown that most cases can be performed without CPB. The question remains as to whether they should be.

There are several limitations to off-pump CABG. It is more difficult to sew on a moving target. Early follow-up data suggest that there is a higher rate of recurrent angina and the need for further intervention in patients done off-pump compared to those done conventionally.[20] The intra-operative stakes are much higher with off-pump surgery. Once one embarks on the off-pump approach, it is hard to bring oneself to put the patient on-pump to finish a difficult lateral wall graft. An emergent need to put patients on CPB cannot be good for them just as much as keeping them off when they should be on. The prospect of reducing the incidence of stroke was an attractive feature of off-pump surgery. The theoretical reason that the stroke incidence was expected to be lower was the avoidance of aorta cross-clamping and the liberation of microemboli caused by the sandblasting effect of the arterial cannula with air bubbles and particles from the pump not caught in the arterial filter. Despite this, the incidence of stroke is similar,[21-22] which could be due to the association of carotid and neurovascular disease in these patients. It is more likely due to the continued requirement for a partial occlusion clamp for the placement of the proximal anastomoses.

Certain groups of patients receive significant peri-operative benefit from off-pump CABG, the elderly, those with calcified aortas, chronic obstructive pulmonary disease, and/or renal failure.[23-26] If the peri-operative benefits are offset by future graft problems, there will be no long-term gain by this approach. The ability to deploy an anastomotic connector rapidly, without error, and reproducibly using an off-pump approach would solve many problems. Employing a no-touch technique to perform the proximal aortic to graft anastomoses could reduce stroke during the off-pump procedure. This background is the impetus for current interest in the development of facilitated anastomotic techniques for coronary artery revascularization.

The Evolution of a Sutureless System for Anastomoses

To address inherent issues in CABG a mechanical device must have quick error free deployment. It must avoid trauma to the intima. To minimize pseudointimal hyperplasia at the anastomosis, it must be composed of inert material. Potential advantages would be the creation of a reproducible, round fit for both proximal and distal anastomoses with the elimination of variations resulting from the suture technique. Grafting would be performed without manipulation of the aorta or the use of CBP. Future applications could include the thoracoscopic placement of bypass grafts or grafting via a percutaneous approach.

A vascular anastomotic device (the Symmetry Bypass System, St. Jude, Inc., Minneapolis, MN), which allows both aorta to vein graft anastomoses and vein graft to coronary artery anastomosis, has been in development for several years. The connectors are made of a nickel titanium alloy (nitinol) or stainless steel. Several generations of the device have undergone extensive experimental testing, and are now being introduced clinically for both proximal and distal anastomoses. This system appears ideal for less invasive coronary revascularization. The connectors allow for intima opposition without everting the vessel or penetrating the wall. Consequently, it can be used on atherosclerotic vessels, in contrast to clip or staple devices, which require eversion. After loading, deployment is rapid and can be performed with minimal training.

The first-generation proximal connector, a nitinol devices allows for aortic to saphenous vein anastomosis in a side-to-end fashion. The saphenous vein is pre-loaded on an introducer and subsequently positioned over the connector. Loading can be accomplished in 1-2 minutes with surgical loupe magnification (2.5 x). The aortotomy necessary for anastomosis is achieved with a unique tool utilizing a round "cookie cutter" type blade which cores out a plug of aorta. This

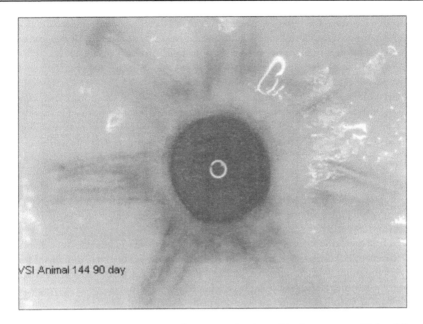

Figure 1. Ostia of an aorto-saphenous vein graft anastomosis using a nitinol connector at 90 days follow-up in a dog model. Photo taken from the aortic side. Note the opaque neointimal covering of the nitinol struts.

technique avoids the crush injury caused to the aortic wall by standard punches. The plug is caught on the aortotomy device by a barb and retracted into the device on its removal. The hole is briefly covered with a finger, and the pre-loaded vein and connector assembly are then slid into the aortotomy site. The anastomosis is created with a movement similar to the click of a ball-point pen, and the introducer is withdrawn. The aortotomy and delivery of the connector are accomplished in < 10 seconds producing a symmetrical anastomosis. Hemostasis is instantaneous.

Experiments have been performed with the nitinol proximal device in dogs. In an acute study, 5 mechanical and 4 sutured aorto-saphenous vein grafts were pressure tested.[27] The pressure causing the first appreciable leak outside connected anastomoses was equal to or greater than that noted with sutured anastomoses. The minimal pressure to any appreciable leak was > 160 mmHg. The tensile force causing an initial leak was > 110 grams for all connected anastomoses. This was also comparable to sutured anastomoses. The proximal connector has been tested experimentally in a chronic model. Eighteen dogs underwent aorto-saphenous vein anastomosis[28] with connectors. Follow-up was 30 days (5 animals), 90 days (8 animals), and 180 days (5 animals). The results were compared to 7 control animals, which received sutured anastomoses. The grafts were widely patent in 100% of studies. Histology revealed very little evidence of inflammation or intimal hyperplasia at the device sites. The connector was partially covered by a thin, translucent layer at 30 days. By 90 days, a thin, opaque layer of neointima covered its surface (Fig. 1) in most cases.

This connector was clinically introduced in Europe in June 1999. It received CE marking in May 2000, and FDA approval in May 2001. Nearly 50,000 proximal anastomoses have been performed. Eckstein et al[29] have reported on 43 consecutive patients undergoing CABG. Of a total of 72 saphenous vein grafts used in these revascularization procedures, 65 were done with a proximal connector. Thirty-eight patients underwent isolated bypass grafting. Twenty-one procedures were performed on-pump and 22 off-pump. Mean diameter of the saphenous vein was 5.0 ± 0.3 mm. The mean loading time was 4.6 ± 1.0 minutes. Mean loading time was considerably shorter as the series progressed. Less than 10 seconds were required for all connections, and hemostasis was absolute in all but 3 subjects. In these cases the device was removed, and the anastomosis sutured to the same aortotomy hole. All grafts were patent, and flows were 50 ± 27 ml/min in single vein grafts, and 83 ± 37 ml/min in sequential grafts. There were no intra- or peri-operative complications related to the device.

Maisano et al[30] have reported a similar experience with 11 off-pump revascularization procedures utilizing aorto-saphenous vein connectors for 17 anastomoses. The indication for the device was a calcified aorta in 6 cases, severe left ventricular dysfunction in 5 cases (to minimize surgical time), and to avoid aortic dissection in 6 patients requiring re-do operations. There was no connector failure and intra-operative flows were excellent. Demertzis and Siclari[31] placed the proximal connector 15 times in 10 patients with 1 failure because of bleeding. In that case, the anastomosis was then sutured without difficulty. While these results are encouraging, long-term patency data remain forthcoming. Indications for clinical use of the device are still to defined. It has been effective in off-pump procedures by avoiding manipulation of the aorta and in patients with a heavily calcified aorta.

The first-generation distal connectors contained significantly less structure than the nitinol connector. They are stainless-steel, catheter-based devices. They were developed for saphenous vein to coronary artery anastomosis, and deployed via a modified Seldinger approach. Because of experimental success, it has been adapted for aorta to saphenous vein graft proximal anastomoses as well. The connector was originally planned for end-to-side anastomosis, but the technique evolved to a side-to-side approach to avoid having to adjust for size differences in various conduits. This approach allows for anastomotic size to equal diameter of the coronary artery target and optimal angulation to avoid kinking of the graft both proximally and distally. The connector contains external fingers to secure the saphenous vein graft, and internal fingers that engage the internal lumen of the artery. These internal fingers are covered by a nose cone to prevent trauma to the coronary artery while the device is being introduced. The connector is mounted on a balloon catheter (Fig. 2A). The balloon expands the connector, creating the anastomosis. At the same time, the connector is reduced in length; compressing the saphenous vein graft to the coronary artery, producing an hemostatic seal with firm attachment of the 2 vessels. To load the vein on the connector, a venotomy is made in the vein wall proximal to the distal end of the graft for placement of the connector. The delivery system is passed through the distal end of the vein and out through the venotomy. The external fingers are then pierced through the vein wall, and a silicone ring covored with titanium dioxide is placed at the level of the external fingers to improve visibility during delivery (Fig. 2B). Loading can be performed on a side table simultaneously with the surgical procedure.

In the chosen coronary artery site, a hole is created with a 20-gauge needle and dilated with a 2.0-mm Teflon-coated dilator. The delivery system is then introduced axially into the coronary artery until the silicone ring is in contact with the coronary artery wall (Fig. 2C). With the device in place, the nose cone is advanced, uncovering the internal fingers, and the delivery system is then positioned at 90° with respect to the coronary artery. This maneuver avoids catching the posterior coronary wall with the internal fingers. The balloon is then inflated to expand the connector (Fig. 2D). Once the connector is expanded, the delivery system is removed, and the vein graft is ligated immediately distal to the anastomosis without compromising the vein-anastomosis lumen. The connector body remains in place and adds structure to the anastomosis (Fig. 2E).

In an experimental study by Schaff et al,[32] 12 dogs underwent on-pump CABG times 2 via a lateral thoracotomy. In each dog, 1 of the 2 distal anastomoses was created with a stainless steel connector (alternating left anterior coronary artery and circumflex coronary artery). The proximals and 2nd distal anastomosis were hand-sewn. At 30-day follow-up, all anastomoses were patent. No significant differences were found when sutured and connected anastomoses

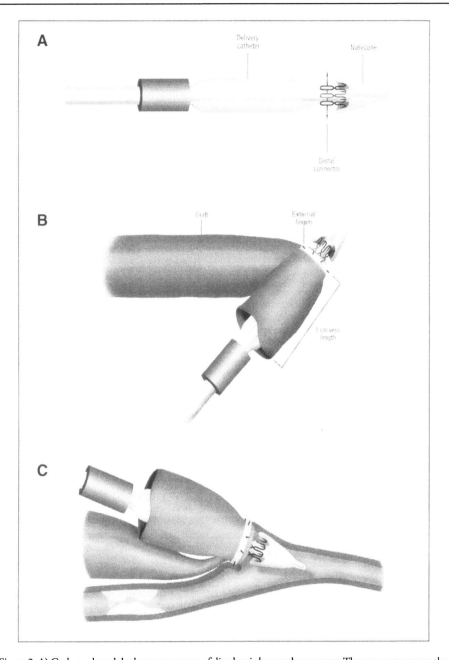

Figure 2. A) Catheter-based deployment system of distal stainless steel connector. The nose cone covers the internal fingers that will be engaged in the coronary artery. The saphenous vein is loaded over the external fingers at right angles to the catheter. B) The saphenous vein after loading on the connector. The external fingers pierce the vein full-thickness. The nose cone covers the internal fingers. C) The loaded saphenous vein connector system is placed into deployment position via a coronary arteriotomy. Figure continued on next page.

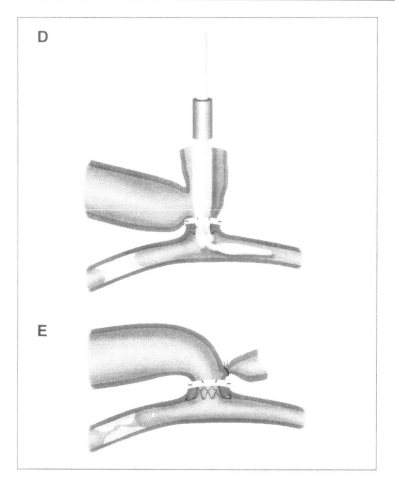

Figure 2, continued. D) The connector is deployed by rolling the nose cone off of the internal fingers and expanding the balloon. E) The finished anastomosis. The distal vein is tied off, effectively creating an end-to-side anastomosis. Reproduced with permission from Annals Thorac Surg 2002; 73:831. ©2002 Society of Thoracic Surgeons.

were compared in regard to intra-operative graft flows, patency, or histology (Fig. 3A and 3B). Mean time for the connector anastomoses was 3 minutes (range 2-5 min) compared to 8.4 min (range 4-13 min) for sutured anastomoses (p < 0.0001). Mean flows determined by Doppler ultrasound were 52.6 ml/min for connected grafts and 59.3 ml/min for sutured grafts after ligation of the native artery proximally (p = 0.4, NS). There were no mechanical failures, and hemostasis was absolute in all anastomoses. Histology at 30 days revealed a thin to moderate amount of fibrous neointima covering the device and fibrous connective tissue surrounding it (Fig. 4). The lumen was not compromised in any case.

Bonilla et al[33] undertook a comparison study to examine burst and tensile strength of the stainless steel distal connector. Anastomoses were tested acutely, and at 30, 90, and 180 days (Fig. 5A and 5B). Comparisons were made between sutured and connected anastomoses. Aorta to coronary artery bypass grafts and femoral to femoral artery saphenous vein bypass grafts were investigated. A total of 34 connected anastomoses and 26 sutured anastomoses were studied. Mean pressure to non-recoverable leak was higher for connectors than for sutures: 471 mmHg

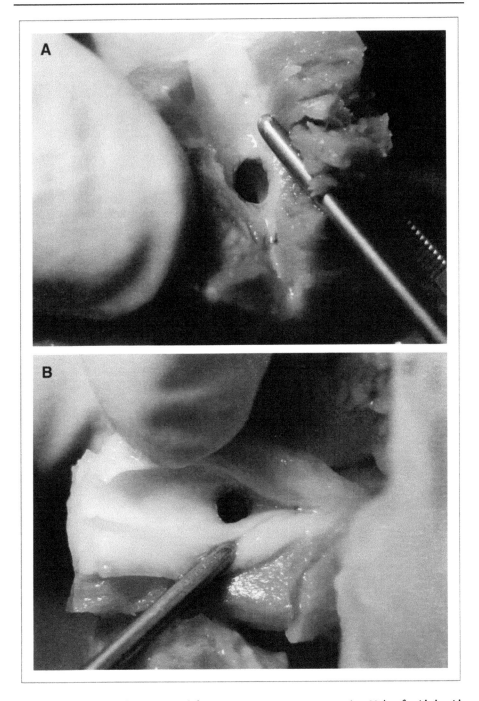

Figure 3. A) A sutured saphenous vein to left anterior coronary artery anastomosis at 30 days for side-by-side comparison with a connected anastomosis (B) in the same animal. Note the suture ligature just proximal to the anastomosis, rendering the coronary artery distribution graft-dependent. B) A connected saphenous vein to left circumflex coronary artery anastomosis at 30 days. Reproduced with permission from Annals Thorac Surg 2002; 73:834. ©2002 Society of Thoracic Surgeons.

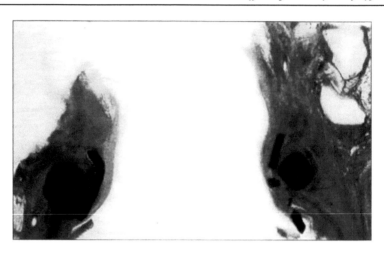

Figure 4. Histology at 30 days of a connected anastomosis. The specimen is cut longitudinally. Note the thin layer of fibrous neointima covering the stainless steel. The lumen is not compromised. Reproduced with permission from Annals Thorac Surg 2002; 73:834. ©2002 Society of Thoracic Surgeons.

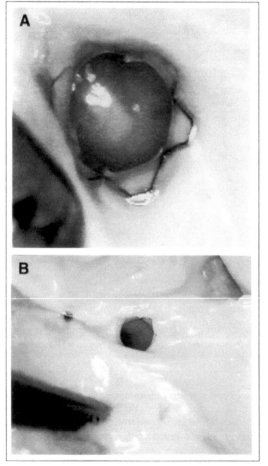

Figure 5. A) A close-up of a connected saphenous vein to left anterior coronary artery anastomosis at 30 days. Photograph taken from the coronary artery side. B) Photograph from the coronary artery side of a typical connected saphenous vein to coronary artery anastomosis at 90 days. Note the neointimal formation without pseudointimal hyperplasia.

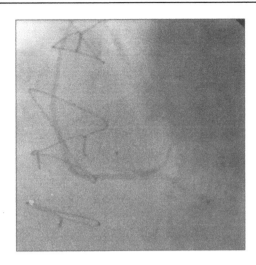

Figure 6. Angiogram taken at 3 months post-operatively of a saphenous vein graft to distal right coronary artery in a patient. Reprinted with permission from The Lancet 2001; 357:931-932. ©2001 Elsevier.

vs. 252.5 mmHg (p < 0.05). Mean tension to non-recoverable leak was higher for sutures than connectors: 0.73 lb vs. 1.92 lb (p < 0.05). All grafts were patent at end-points. Mean time to suture was 10.9 min compared to 2.6 min for connected anastomoses (p < 0.05). Mean flow was 83 ml/min for connected grafts, and 45.3 ml/min for sutured ones (p < 0.05). In this study, connected anastomoses were more effective than suture anastomoses in regard to handling pressure without leak. They did not tolerate being pulled on as much as sutured anastomoses. However, tensile strength was more than adequate. These data supported proceeding to clinical application.

The first-generation distal device was introduced clinically by Eckstein et al,[34] who have reported recently the first clinical deployment in the *Lancet*. A saphenous vein graft was anastomosed to the distal right coronary artery in a 61-year-old male with 3-vessel coronary artery disease. Flow was 80 ml/min, and angiography at 3 months post-operatively revealed a widely patent graft (Fig. 6). A total of 14 distal anastomoses were performed with excellent results. However, the loading procedure on the back table proved cumbersome and was time-consuming. Ongoing development led to a second- generation, easy-load system, which allows loading at the chest after the proximal anastomosis has been placed.

The second-generation distal device is similar to the first-generation device. However, the act of passing the catheter-loaded device through the venotomy is all the loading required (Fig. 7A and 7B). This has dramatically changed the procedure. It allows the proximals to be done first, followed by the distals. Earlier experimental work has shown the feasibility of this system for the creation of both proximal and distal anastomoses in the same graft (Fig. 8) in an on-pump porcine model under fibrillatory arrest, using proximal and distal stainless steel connectors.[35] Each graft was placed in < 3 min. Hemostasis was immediate and absolute. Development of the easy-load device was the key to clinical application.

The second-generation distal device has been studied experimentally in both acute off-pump and chronic on-pump canine model.[36] A single saphenous vein graft-to-left anterior descending (LAD) artery procedure (n=18) was performed to evaluate long-term patency of the 2-mm ID stainless steel distal connector in the chronic canine model. All grafts were widely patent after a minimum 30 (n=8) (Fig. 9A and 9B), 90 (n=5) (Fig. 10A and 10B), and 180 days (n=5) (Fig. 11A and 11B) in this model. An acute off-pump feasibility study was undertaken on saphenous vein graft-to-LAD bypass fashioned with both proximal and distal connectors (n=15) total grafting time, including loading and deployment, was 10:54 ± 2:54 min.

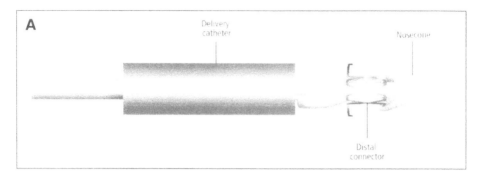

Figure 7A. Second generation distal connector and delivery system. Reprinted with permission from EJCTS 2003; 23:926. ©2003 Elsevier.

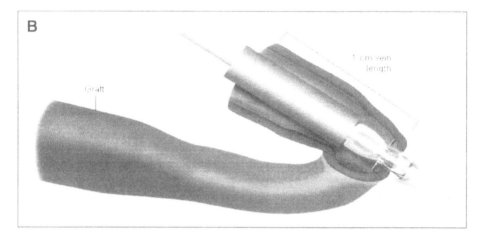

Figure 7B. Connector and delivery system loaded into the graft. Reprinted with permission from EJCTS 2003; 23:926. ©2003 Elsevier.

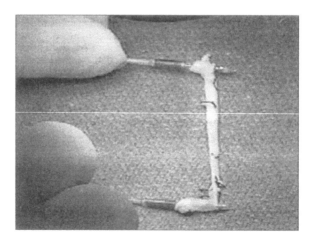

Figure 8. Saphenous vein loaded for both proximal and distal connector deployment in an animal experiment.

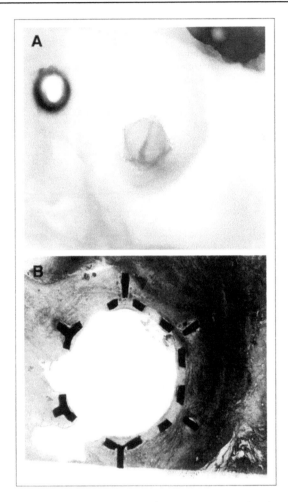

Figure 9. A) Gross appearance of distal device-created anastomosis at 30 days (magnification x 4). B) Microscopic appearance of distal device-created anastomosis at 30 days (magnification x 40). Reprinted with permission from EJCTS 2003; 23:932. ©2003 Elsevier.

Subsequently, the second-generation device was introduced in Europe. Although there has been significant experimental experience and now early clinical experience with the distal device, laboratory work with atherosclerotic vessels has been limited. Appropriate indications and contraindications related to clinical application will require more experience. To be maximally effective, the devices will need to be applicable to ITA to coronary artery anastomoses.

Current Status of Other Facilitated Techniques

Other groups have shown considerable interest in facilitated anastomoses for coronary artery revascularization. Their reasons are similar: to avoid aortic manipulation and cut down on the time to create a distal anastomosis to enable off-pump surgery. Many devices are in experimental development, and several have been introduced clinically in a limited manner for proximal anastomoses.

Kirsch et al[37] developed a non-penetrating microvascular clip to create an anastomosis by intimal apposition in an interrupted technique. This technique has been successfully used for

Figure 10. A) Gross appearance of distal device-created anastomosis at 90 days (magnification x 4). B) microscopic appearance of distal device-created anastomosis at 90 days (magnification x 40). Reprinted with permission from EJCTS 2003; 23:932. ©2003 Elsevier.

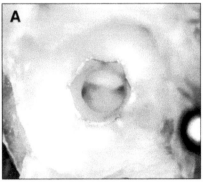

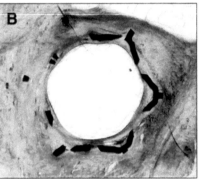

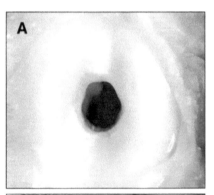

Figure 11. A) Gross appearance of distal device-created anastomosis at 180 days. (magnification x 4). B) appearance of distal device-created anastomosis at 180 days (magnification x 40). Reprinted with permission from EJCTS 2003; 23:932. ©2003 Elsevier.

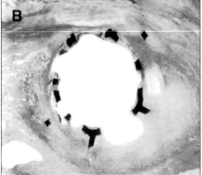

coronary anastomosis, which heals well with minimal fibrosis and intimal hyperplasia. Although faster than suturing, the main drawback is the requirement for an eversion technique. Thus, their use in atherosclerotic vessels is limited. It is doubtful that the clips can be placed easily on a beating heart in off-pump surgery. Further development of this concept has resulted in a device that lays down the entire clipped anastomosis with "One-shot" (US Surgical Corporation, Norwalk, CT). There has been considerable experimental testing of both proximal and distal vein graft to coronary artery anastomoses. In a porcine study, 14 out of 14 anastomoses were patent at 1 month post-operatively.[38] However, 43% were suboptimal. The device was approved for clinical use, but has thus far failed to enjoy widespread application.

A non-suture alternative has been developed recently. The nitinol coil device (U-Clip, Coalescent Surgical, Sunnyvale, CA) is available clinically for conventional suturing while avoiding knot-tying. It is placed in an interrupted fashion with suture needles. After positioning, the deployment wire is detached, and the nitinol portion seeks its coil shape, apposing the vein to the coronary artery. Ono et al[39] have reported on 13 patients in whom the suture-clips were used for the left ITA to LAD coronary artery anastomosis. Four patients underwent minimally invasive direct coronary artery bypass (MIDCAB), 5 OPCAB, and 4 conventional CABG. A mean of 11.8 clips (range 9-15) were placed and the mean time to anastomosis creation was 15.9 min (range 10-30 min). All patients did well. Patency was confirmed 6 months post-operatively in all subjects. According to the information released by Coalescent Surgical, more than 1,000 proximal and distal anastomoses have been performed clinically.[40] A mean number of 10.8 clips were used for anastomoses requiring a mean time of 8.5 min. This device does not result in substantial time reduction compared to suturing. It does not allow avoidance of an aortic partial occlusion clamp to create proximal anastomoses. The device may be applicable to a port access approach for robotic creation of distal anastomoses where it may be advantageous to avoid knot-tying.

Several devices, currently in development, are driven by the concept of a semi-automated connector to facilitate proximal aortic to conduit anastomosis. Calafiore et al[41] have reported their successful experimental and clinical experience with a nitinol connector device for proximal anastomoses (Aortic Anastomotic Device, Cardiovations, Ethicon, Inc., a Johnson and Johnson Company, Sommerville, NJ). Thirty-two connector proximal anastomoses were performed in 16 sheep. They were compared with sutured anastomoses. Patency for connector anastomoses was 87.5% compared to 85% for sutured anastomoses. The anastomoses were rapidly created (1-2 minutes). The device was used in 11 patients with only 1 problem anastomosis. Clinical trials are ongoing.

Hausen et al[42] have described a novel, integrated, automatic system which allows proximal aorta to saphenous vein graft anastomoses to be placed in less than 5 sec in sheep. Seven connected anastomoses were compared to 6 hand-sewn controls. At 9-week follow-up, 5 of 6 hand-sewn and 6 of 7 connected grafts were patent. All occlusions were related to distal hand-sewn anastomosis. The average diameter of the connected anastomoses was 4.0 ± 0.2 mm compared to 3.1 ± 1.7 mm when hand-sewn.

There are few reported successes with connector devices to perform distal conduit to coronary artery anastomoses. Adams et al[43] undertook an experimental study of a novel vascular coupling system (Magnetic Vascular Positioner, MVR, Ventrica, Menlo Park, CA). Right ITA to right CABG were performed in an off-pump porcine model with pairs of elliptical magnetic implants. One pair created the coronary artery-docking port, the other a graft-docking port. The ports were coupled magnetically. Anastomoses were successful in 19 animals without the need for a stabilization device. All grafts in the surviving animals were patent at 30 days. Patency was demonstrated at 90 days in selected animals.

Solem et al[44] introduced a T-shaped nitinol connector device with a Teflon (PTFE) covering (Solem GraftConnector, Jomed International, Helsingborg, Sweden) to create ITA to LAD coronary artery anastomoses. The ITA is loaded into the connector. The T-shaped connector is then inserted into a 7 mm arteriotomy in the LAD coronary artery. After release, the self-expanding stent anchors the end-to-side anastomosis without any necessary suturing. Eleven connector anastomoses were compared with 14 sutured anastomoses in a beating heart coronary artery bypass model in sheep. The connector anastomoses were performed in a mean time of 2.4 ± 0.2 min compared to 6.9 ± 0.4 min ($p < 0.0001$) for sutured anastomosis. The native vessels were ligated proximally, and graft flows were similar in each group.

Comments

The last 100 years has seen considerable trial and error regarding facilitated anastomotic techniques. Remarkably rapid progress has been made over the last 5 years in the development of a rapidly-deployed connector system for coronary artery revascularization. This has been largely spurred by a vision based on theoretical possibilities. Proximals can be attached with a no-touch technique. This can benefit both on- and off-pump coronary revascularization. In patients with heavily calcified aortas, there is often a small soft area that can be cannulated for perfusion or where a proximal can be placed. A no-touch technique would allow this conduit to be located on the aorta, but still avoid cross-clamping. The no-touch technique could reduce complications in other revascularization procedures, such as aorto-renal artery reconstructions, and lower extremity procedures. Distal anastomoses could be performed in a fraction of the time that it takes to sew a graft in place. This would reduce CPB time in on-pump procedures, and would markedly decrease the problem with hemodynamic compromise in patients undergoing off-pump revascularization. It would be particularly useful in patients not tolerating the manipulations necessary to graft lateral wall vessels in off-pump procedures. Current delivery systems could be adapted to allow for graft placement from a distance. One could envision a left-sided vein graft positioned via a thoracoscopic port. The ITA could be placed similarly after take-down from the chest wall, using robotic endoscopic technology. Perhaps one day, a clinical CABG may be installed percutaneously. Once feasibility has been demonstrated experimentally, adaptation often allows for an appropriate clinical niche. Indeed, this percutaneous maneuver has been performed successfully in a porcine model.[45] A saphenous vein was introduced into the aorta through the femoral artery and ultimately passed back out through the ascending aorta, then subsequently connected to the left anterior ascending coronary artery (Fig. 12). Although one can think of a myriad of problems that could be associated with this approach, when vision exists, success often follows.

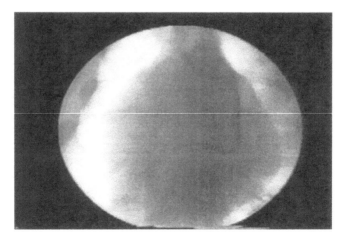

Figure 12. Angiogram of a saphenous vein graft from the aorta to the left anterior coronary artery placed percutaneously via the common femoral artery in an experimental procedure.

Success with connectors could have an enormous impact on coronary revascularization. The reality will likely be somewhere in between. Even though the prospect of performing an anastomosis in sec to min is appealing, acceptance will be slow because of the proven long-term patency associated with sutured CABGs placed with the benefit of CPB and cardioplegia. The need for long-term patency data will be necessary. Experimental and clinical data remain encouraging.

References

1. Carrel A. La technique operatoire des anastomoses vasculaires et la transplantation des visceres. Lyon Med 1902; 98:859-63.
2. Androsov PI. New method of surgical treatment of blood vessel lesions. AMA Arch Surg 1956; 73:902-10.
3. Inokuchi K. Stapling device for end-to-side anastomosis of blood vessels. Arch Surg 1961; 82:337-41.
4. Nakayama K, Tamiya T, Yamamoto K et al. A simple new apparatus for small vessel anastomoses. Surgery 1962; 52:918-23.
5. Ostrup LT, Berggren A. The Unilink instrument system for fast and safe microvascular anastomosis. Ann Plast Surg 1986; 17:521-5.
6. Nataf P, Hinchliffe P, Manzo S et al. Facilitated vascular anastomoses: the one-shot device. Ann Thorac Surg 1998; 66:1041-4.
7. Payr E. Zur frage der circularen vereinigung von blutgefasse mit resorbirbaren prothesen. Arch Klin Chir 1904; 72:32-54.
8. Carter EL, Roth EJ. Direct non-suture coronary anastomoses in the dog. Ann Surg 1958; 148:212-8.
9. Rohman M, Goetz RH, Dee R. Double coronary artery internal mammary artery anastomoses: tantalum ring technique. Surg Forum 1960; 11:236-7.
10. Ratan RS, Leon M, Lovette JB et al. Modified non-suture anastomosis of coronary artery and internal mammary artery in dogs. Surg Forum 1960; 11:239-41.
11. Holt GP, Lewis FJ. A new technique for end-to-end anastomosis of small arteries. Surg Forum 1960; 11:242-3.
12. Haller JD, Kripke DC, Rosenak SS et al. Long-term results of small vessel anastomoses with a ring technique. Ann Surg 1965; 161:67-72.
13. Demikhov VP. Experimental transplantation of vital organs. Authorized translation from the Russian by Basil Haigh. New York: Consultant's Bureau, 1962.
14. Murray G, Hilario J, Porcheron R et al. Surgery of coronary heart disease. Angiology 1953; 4:526-31.
15. Konstantinov I. Robert H. Goetz: the surgeon who performed the first successful clinical coronary artery bypass operation. Ann Thorac Surg 2000; 69:1966-72.
16. Kolesov VI. Mammary artery-coronary artery anastomosis as a method of treatment for angina pectoris. J Thorac Cardiovasc Surg 1967; 54:535-44.
17. Buffolo E, de Andrade CS, Branco JN et al. Coronary artery bypass grafting without cardiopulmonary bypass. Ann Thorac Surg 1996; 61:63-6.
18. Benetti FJ, Mariani MA, Ballester C. Direct coronary surgery without cardiopulmonary bypass in acute myocardial infarction. J Cardiovasc Surg (Torino) 1996; 37:391-5.
19. Tasdemir O, Vural KM, Karagoz H et al. Coronary artery bypass grafting on the beating heart without the use of extracorporeal circulation: review of 2052 cases. J Thorac Cardiovasc Surg 1998; 116:68-73.
20. Arom KV, Flavin TF, Emery RW et al. Safety and efficacy of off-pump coronary artery bypass grafting. Ann Thorac Surg 2000; 69(3):704-10
21. Kshettry VR, Flavin TF, Emery RW et al. Does multivessel, off-pump coronary artery bypass reduce postoperative morbidity? Ann Thorac Surg 2000; 69:1725-30.
22. Hernandez F, Cohn WE, Baribeau YR et al. In-hospital outcomes of off-pump versus on-pump coronary artery bypass procedures: a multicenter experience. Ann Thorac Surg 2001 72(5):1528-33.
23. Petro KR, Dullum MK, Garcia JM et al. Minimally invasive coronary revascularization in women: a safe approach for a high-risk group. Heart Surg Forum 2000; 3:41-6.
24. Yokoyama T, Baumgartner FJ, Gheissari A et al. Off-pump versus on-pump coronary bypass in high-risk subgroups. Ann Thorac Surg 2000; 70(5):1546-50.

25. Boyd WD, Desai ND, Del Rizzo DF et al. Off-pump surgery decreases postoperative complications and resource utilization in the elderly. Ann Thorac Surg 1999; 68(4):1490-3.

26. Ascione R, Lloyd CT, Underwood MJ et al. On-pump versus off-pump coronary revascularization: evaluation of renal function. Ann Thorac Surg 1999; 68(2):493-8.

27. Berg T, Bonilla LF. Anastomotic Technology Group, St. Jude Medical Inc., Minneapolis, MN, U.S.A., Unpublished data, Personal communication.

28. Berg T, Bonilla LF. Anastomotic Technology Group, St. Jude Medical Inc., Minneapolis, MN, U.S.A., Unpublished data, Personal communication.

29. Eckstein FS, Bonilla LF, Englberger L et al. The St Jude Medical symmetry aortic connector system for proximal vein graft anastomoses in coronary artery bypass grafting. J Thorac Cardiovasc Surg 2002; 123(4):777-82.

30. Maisano F, De Bonis M, Greco P et al. Anatomotic devices and beating heart surgery: a preliminary experience. Presented at the 4th Annual Meeting of the International Society for Minimally Invasive Cardiac Surgery, June 27-30, 2001, Munich, Germany.

31. Demertzis S, Siclari F. Initial experience with the aortic connector device for sutureless proximal vein graft anastomoses without aortic clamping. Presented at the 4th Annual Meeting of the International Society for Minimally Invasive Cardiac Surgery, June 27-30, 2001, Munich, Germany.

32. Schaff HV, Zehr KJ, Bonilla LF et al. An experimental model of saphenous vein-to-coronary artery anastomosis with the St. Jude Medical stainless steel connector. Ann Thorac Surg 2002r; 73(3):830-5.

33. Bonilla LF, Bianco RW, Berg T et al. A novel sutureless mechanical device for autologous vein to coronary artery anastomoses: feasibility in the dog model. University of Minnesota, Minneapolis, MN, U.SA., Unpublished data. Personal communication.

34. Eckstein FS, Bonilla LF, Meyer B et al. Sutureless mechanical anastomosis of a saphenous vein graft to a coronary artery with a new connector device. Lancet 2001; 357:931-2.

35. Zehr KJ, Schaff HV, Bonilla LF et al. Sutureless coronary artery bypass grafting. Presented at the 7th Annual Cardiothoracic Techniques and Technologies Meeting, January 24-27, 2001, New Orleans, LA; U.S.A.

36. Zehr KJ, Hamner C, Bonilla LF et al. Evaluation of a novel 2.0 mm ID stainless steel saphenous vein to coronary artery connector: laboratory studies of on-pump and off-pump revascularization. Presented at the 16th annual scientific sessions of the European Association for Cardio-Thoracic Surgery, September 22-25, 2002, Monte Carlo, Monaco.

37. Kirsch WM, Zhu YH, Hardesty RA et al. A new method for microvascular anastomosis. Am Surg 1992; 58:722-7.

38. Heijmen RH, Hinchliffe P, Borst C et al. A novel one-shot anastomotic stapler prototype for coronary bypass grafting on the beating heart: feasibility in the pig. J Thorac Cardiovasc Surg 1999; 117:117-25.

39. Ono M, Wolf RK, Angoulas D et al. Clinical trial of LIMA-to-LAD anastomosis using new suture-clip device. Presented at the 4th Annual Meeting for the International Society for Minimally Invasive Cardiac Surgery, June 27-30, 2001, Munich, Germany.

40. Hill A. Coalescent nitinol suture clip for distal coronary artery. Presented at the 7th Annual Cardiothoracic Techniques and Technologies Meeting, January 24-27, 2001, New Orleans, LA, U.S.A.

41. Calafiore AM, Barel Y, Vitolla G et al. Animal and clinical study on a new sutureless anastomotic device for the proximal SVG anastomosis. Presented at the 80th Annual Meeting of the American Association of Thoracic Surgery, April-May, 2000, Toronto, Canada.

42. Hausen B, Gruenenfelder J, Yencho SA et al. Rapid deployment of a vein to aorta anastomosis with a novel integrated and automated anastomosis system: preclinical results from an aorto-coronary bypass study in sheep. Circulation 2000; 102(Suppl II):II-766.

43. Adams D, Filsoufi F, Farivar RS et al. Sutureless distal coronary bypass using a novel magnetic coupler. Presented at the 4th Annual Meeting of the International Society for Minimally Invasive Cardiac Surgery, June 27-30, 2001, Munich, Germany.

44. Solem JO, Boumzebra D, Al-Buraiki J et al. Evaluation of a new device for quick sutureless coronary artery anastomosis in surviving sheep. Eur J Cardiothorac Surg 2000; 17:312-8.

45. Berg T, Swanson W, Hindrichs P. Percutaneous aorta to coronary artery saphenous vein bypass grafting. Unpublished data. Personal communication.

CHAPTER 14

The Future of Coronary Artery Surgery:

Quo Vadimus?

Ray C.-J. Chiu

As Dr. Paul Cartier vividly described in the Foreword of this book, there has been great progress in coronary artery surgery in the past half century. Thanks to the courageous and ingenious efforts of pioneers like him, coronary artery bypass today is a routine procedure, one of the most common major operations in developed countries. The magnitude of advancement has been so spectacular that new surgical concepts diametrically opposed to those widely accepted a generation ago are being taught to our young surgeons. Only a decade or two ago, they were trained to make incisions large enough to obtain adequate surgical exposure, and not to struggle through small openings, since "wounds heal from side-to-side and not from end-to-end!". Today, minimally invasive surgery is in vogue, and they are encouraged to accomplish the procedure with minimal incision. Likewise, it was taken for granted that delicate cardiac surgery could be accomplished well only in a "quiet and bloodless operative field"; thus cardiopulmonary bypass and cardioplegia were believed to be mandatory. Today, as this book illustrates, off-pump bypass is becoming a common and preferable approach in an increasing number of patients.

Compared to a generation ago, we have reached a new level of sophistication and success, and yet we must ask where we are going with this procedure. Predicting the future is always fraught with danger, but we can be certain about one thing coronary artery surgery will not remain the same. Sometimes, a major quantum leap in knowledge or technology occurs, resulting in unpredictable future paradigm shifts. But one may still peek into the future by extrapolating current developments. Perhaps we could postulate three scenarios that may develop in sequence or in an overlapping and interacting manner.

The first scenario, namely, '*competition between coronary artery stenting and coronary artery bypass surgery*', is already underway, as newly-developed, pharmacologically-coated intravascular stents appear to be more resistant to restenosis.[1] In many centers, a trend of reduction in the number of cases for coronary bypass surgery has already been discerned.[2] The future choice between these two approaches, i.e., stent versus bypass, will largely depend on three factors: the durability of patency and the incidence of restenosis; the magnitude of invasiveness for the procedure; and the question of cost under the current health care delivery system. At present, stents are considered to be more restenosis-prone, but less invasive for their insertion, and less costly, compared to surgical revascularization procedures.[3] Thus, newer surgical techniques, less invasive and not as expensive, would make the surgical approach more competitive. The off-pump bypass described in this book eliminates the cost of a pump oxygenator, and reduces the morbidity caused by hematological and inflammatory responses associated with extra-corporeal circulation. Since both the stenting and surgical techniques for coronary

---

revascularization are making continuous progress, the final outcome of this issue will not be determined in the immediate future.

Another sphere of development, which could profoundly affect the future of cardiac surgery in general, is associated with progress '*from minimally invasive operation to robotic surgery*'. At present, robotic surgery is in its infancy, and looks like an expensive and cumbersome exercise.[4] Nevertheless, with continued advances, combining increasingly sophisticated effector mechanics with imaging and computing technologies, a major revolution in surgical techniques is conceivable. Mechanical manipulators have made progress in recent years to facilitate highly-delicate operative maneuvers while filtering out undesired vibrations and tremors. Active research is on-going to incorporate "haptic feedback", enabling the surgeon to regain his or her tactile sensation, an important sensory input for delicate surgical procedures. Without such sensation, even tying a fine suture can become a challenging task. The term "virtual fixture" represents a preprogrammed series of maneuvers in which defined operative procedures, such as tying a knot are automated, thus making them uniform, rapid and secure. Ultimately, interfacing of on-line imaging, deploying technologies such as super-high speed nuclear magnetic resonance imaging (NMR) or 3-D echo, can capture the anatomical images during surgery. Then, such images could be used to generate computer-aided-design software, which could guide mechanical effectors to automate the surgical procedure under the supervisory eyes of a surgeon.[5] If this scenario sounds incredible, imagine Charles Lindbergh, who pioneered the first solo flight in a single engine airplane, St. Louis, from America to Paris, being told that one day such a flight would be carried out routinely every day by equipment known as an "autopilot"!

Another development, which may impact on future coronary artery surgery, is '*angiogenic therapy*'. This approach is already undergoing preclinical and clinical trials. By delivering angiogenic factor proteins such as bFGF, VEGF, or the genes encoding them, hopefully new blood vessels will be regenerated in the ischemic myocardial segments supplied by occluded coronary arteries.[6,7] More recently, circulating vascular progenitor cells and bone marrow-derived stem cells have been studied to grow new blood vessels, a process known as "vasculogenesis".[8] The clinical trial results obtained so far have been inconsistent, and this may be due in part to an excessively simple approach of delivering a single angiogenic factor, such as VEGF, for an arbitrary period of time, which is inconsistent with the normal physiology of angiogenesis. It is now known that a series of angiogenic factors and their receptors are involved in a sequential or synergistic manner, not only to produce primitive vessels, but also to guide them to maturity and remodeling, resulting in a vascular network for effective tissue perfusion. Much more information on how these signaling molecules are orchestrated, and the role of mechanical remodeling in response to blood flow shear stress, etc., is required to attain our therapeutic goal in the future. Furthermore, these new vascular networks need to have unobstructed feeding vessels, which are often epicardial coronary arteries prone to arteriosclerotic occlusion. Thus, in the short-run, angiogenic treatment may be particularly useful for patients with diffuse coronary arteriosclerosis and poor run-off, so that the coronary artery bypass procedure could be combined with angiogenic therapy to improve both tissue perfusion and graft patency.

The ultimate answer to the question of how to best manage coronary artery disease is the prevention of coronary arteriosclerosis, and the therapeutic regression of occlusive coronary lesions. When this Holy Grail is attained, coronary artery surgery will have accomplished its mission, and cardiac surgeons of the future will have to turn their attention to other challenges.

References

1. Pearce BJ, McKinsey JF. Current status of intravascular stents as delivery devices to prevent restenosis. Vasc Endovascular Surg 2003; 37(4):231-7; discussion 237.
2. Poyen V, Silvestri M, Labrunie P et al. Indications of coronary angioplasty and stenting in 2003: what is left to surgery? J Cardiovasc Surg (Torino) 2003; 44(3):307-12.
3. Cohen DJ, Breall JA, Ho KK et al. Economics of elective coronary revascularization. Comparison of costs and charges for conventional angioplasty, directional atherectomy, stenting and bypass surgery. J Am Coll Cardiol 1993; 22(4):1052-9.
4. De Cannierere D. Closed chest coronary surgery. State of the art. J Cardiovasc Surg (Torino) 2003; 44(3):323-30.
5. Boyd WD, Kodera K, Stahl KD et al. Current status and future directions in computer-enhanced video- and robotic-assisted coronary bypass surgery. Semin Thorac Cardiovasc Surg 2002; 14(1):101-9.
6. Sakakibara Y, Tambara K, Sakaguchi G et al. Toward surgical angiogenesis using slow-released basic fibroblast growth factor. Eur J Cardiothorac Surg 2003; 24(1):105-12.
7. Nagy JA, Dvorak AM, Dvorak HF. VEGF-A(164/165) and PlGF: Roles in angiogenesis and arteriogenesis. Trends Cardiovasc Med 2003; 13(5):169-75.
8. Chiu RC. Bone-marrow stem cells as a source for cell therapy. Heart Fail Rev 2003; 8(3):247-51.

INDEX

T - #0145 - 111024 - C170 - 229/152/8 - PB - 9780367446512 - Gloss Lamination